Petromania

Black gold, paper barrels and oil price bubbles

by Daniel O'Sullivan

HARRIMAN HOUSE LTD

3A Penns Road
Petersfield
Hampshire
GU32 2EW
GREAT BRITAIN

Tel: +44 (0)1730 233870
Fax: +44 (0)1730 233880
Email: enquiries@harriman-house.com
Website: www.harriman-house.com

First published in Great Britain in 2009

Copyright © Harriman House Ltd

The right of Daniel O'Sullivan to be identified as the author has been asserted in accordance with the Copyright, Design and Patents Act 1988.

ISBN: 978-1-906659-24-0

British Library Cataloguing in Publication Data
A CIP catalogue record for this book can be obtained from the British Library.

All rights reserved; no part of this publication may be reproduced, stored in a retrieval system, or transmitted in any form or by any means, electronic, mechanical, photocopying, recording, or otherwise without the prior written permission of the Publisher. This book may not be lent, resold, hired out or otherwise disposed of by way of trade in any form of binding or cover other than that in which it is published without the prior written consent of the Publisher.

Printed and bound in the UK by CPI, Antony Rowe, Chippenham

No responsibility for loss occasioned to any person or corporate body acting or refraining to act as a result of reading material in this book can be accepted by the Publisher, by the Author, or by the employer of the Author.

To Ele and Nessa, with love

Contents

Table of Figures — vii

Acknowledgements — ix

Preface — xi

Introduction — 1

1. The Ascent — 13
2. Paper Barrels — 49
3. The Financialisation of Oil — 87
4. The Peak Weeks — 131
5. A Bubble by Any Other Name — 163
6. *Petromania Redux* — 213

Sources & Bibliography — 253

Index — 277

Table of Figures

Figure 1: 21st century oil prices, the story so far — 4

Figure 2: Two bubbles — 6

Figure 3: The ascent to oil's peak — 41

Figure 4: Sample oil futures curves — 55

Figure 5: Growth of Nymex open interest — 70

Figure 6: Nymex oil market growth by CFTC "commercial"/ "non-commercial" definitions — 98-99

Figure 7: Roll yields on futures curves in contango and backwardation — 112

Figure 8: Nymex oil market growth with more detailed trader definition — 123-4

Figure 9: Changing trader group domination across maturities from 2000-2008 — 152-3

Figure 10: Mad May — 154

Figure 11: Forces bearing on oil price formation in the Nymex futures market — 188

Figure 12: Long-term price graph comparison for Nymex oil and Nasdaq 100 — 209

Figure 13: Two bubbles revisited – and what the Nasdaq did next — 211

Figure 14: Charting a fair price for oil — 249

Acknowledgements

Many oil industry experts are mentioned by name somewhere in this book, and it will become apparent that I agree with some and disagree with others. I would like to stress, however, that where there is such disagreement on my part, it is above all respectful disagreement. In this spirit I would like to thank all of the following people in particular for their freely-given time in discussing various issues with me, whether personally or in correspondence, whether in specific regard to the research for this book or in my day-to-day job as a journalist: Ed Morse, Daniel Ahn, Jeffrey Currie, Francisco Blanch, Colin Smith, Stephen Schork, Robert McCullough Jr., Christof Ruehl and Roger Bentley.

Very special thanks are due to Leo Drollas of the Centre for Global Energy Studies – unwittingly and unbeknownst to him (until now!) he is in a certain sense the godfather to this book. It was after a couple of hours spent in conversation with him in his London offices in January 2009 that I finally decided there was enough of a disjuncture between what actually happened with the oil price in 2008, and what an establishment consensus still persisted in saying had happened, to justify treatment of these issues at greater length. Mr Drollas himself is certainly not part of this establishment consensus, rather a proud outlier; and although I have not quoted him extensively in this work, this should in no way detract from the debt I owe him and the fact that it was he who first pointed me toward an extremely wide range of evidence backing up a stance I had already taken in early 2008 on a then significantly narrower base of evidence, bolstered by inference.

Speaking of my day-to-day job, I would also like to acknowledge various editorial colleagues at the *Investors Chronicle* who have taken charge of our news coverage from time to time over the past few years, all of whom have proved happy in that role to support and run with my own take on the oil markets. When we originally called the oil price as a bubble in May 2008, this placed us well outside the then-fashionable groupthink dominating the financial press. Thanks in this regard are owed to Graeme Davies, Oliver Ralph, Simon Thompson, John Hughman and Jonathan Eley.

Paul J. Davies and Chris Dillow were both kind enough to read and provide invaluable criticism, comment, and friendly encouragement on vast tracts of the manuscript as it emerged, and Andrew Adamson and Jack Cross also provided crucial help with the drafting. I am grateful to the team at Harriman House, particularly Stephen Eckett, Chris Parker and Suzanne Anderson, for taking this project on and seeing it through to completion in a very tight timeframe, in a very professional manner, yet in an ever-amiable and relaxed fashion throughout. It goes without saying that any simplifications, exaggerations, omissions and other errors still populating this text are my responsibility alone.

Last but certainly not least, it would never have been possible for me to steal the time in which to produce this book without a great deal of love, support and forbearance shown me by the two beautiful girls in my life, my wife Eleanor and daughter Nessa – to whom this book is dedicated, and for whom I promise to try to be (slightly) less boring and obsessional in future.

Daniel O'Sullivan

London, 2009

Preface

Petromania is an account of the spectacular boom and bust that occurred in the crude oil market through 2008, which saw oil breach $100 per barrel at the start of the year, clock an as-yet all-time high of $147 per barrel in July, but then collapse to $34 per barrel by Christmas – the wildest price movement ever seen in our most important global commodity, and one which wreaked havoc in the worldwide economy.

Contrary to arguments advanced by many commentators focusing on fundamental factors, this book argues that new forms of financial speculation which sprung up in commodity markets from the 1990s onwards were instead the key driving force behind this chain of events, now positively identified as a classic speculative bubble. No prior knowledge of the oil markets and their functioning is presumed, so it is hoped this text may also profitably serve as a basic primer on these subjects as well.

I have for the most part eschewed footnotes, but all of the key sources and texts can be found in the Sources & Bibliography section at the back of the book.

Introduction

A Shadowy History

2008 was the year the oil price made history not just once but three times. The world gasped as it breached the $100 per barrel threshold, for the first time ever, on the very first trading day of that year. We swooned as crude oil reached an all-time high, as yet unsurpassed, of $147 per barrel in early July. And when oil subsequently collapsed from that peak in its most precipitate dollar decline ever, to trade at $34 per barrel by Christmas '08, people did not know whether to laugh or cry. Had all that suffering been for naught? Because make no mistake, no price matters more in this world than that of a barrel of crude oil. Along its way, the rampaging oil price had played havoc with livelihoods worldwide – affecting the poorest people in particular – whilst crippling and bankrupting businesses across the globe, destabilising developing world governments, and preventing developed world governments from cutting interest rates as swiftly as needed in meeting the most serious international economic crisis since the Great Depression.

Figure 1 shows how extraordinary this price blow-out was in terms of what has gone before since the turn of the 21st century (through the 1990s prices were even lower, around $20 per barrel). For a physical commodity such as crude oil there are always physical market fundamentals to consider in any account of price movements. And as we shall see, there were many and indeed still remain many who feel that supply and demand fundamentals justified the wild appreciation in oil prices seen in recent years. Yet whenever such a spectacular boom and bust in any asset price is observed over such a short timeframe, an inescapable suspicion is that it was above all the product of a "speculative bubble" inflated by investors – as defined by Yale economist Robert Shiller, 'a situation in which temporarily high prices are sustained largely by investors' enthusiasm rather than by consistent estimation of real value.'[1]

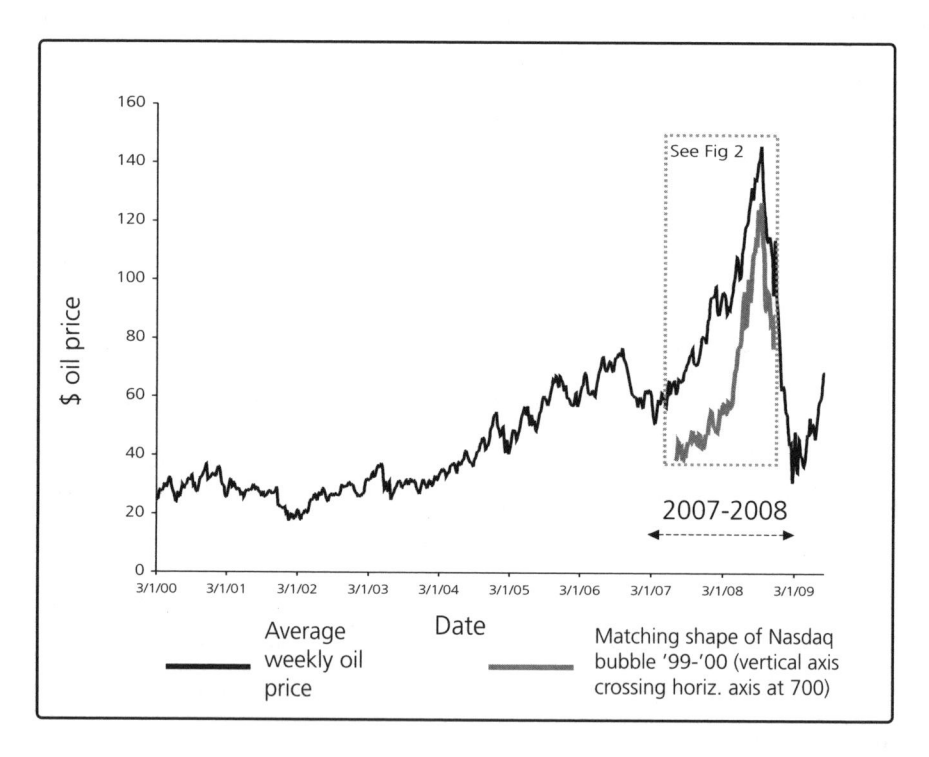

Figure 1: 21st century oil prices, the story so far   [Source: Thomson Datastream]

This book is overwhelmingly concerned with establishing that the actual physical fundamentals of oil supply and demand patently failed to justify crude oil prices at the levels they achieved in summer 2008, but that activity on the part of speculative financial investors was instead the major driving force behind this phenomenon. The arguments entertained both for and against this proposition will range

---
[1] *Irrational Exuberance*, Robert Shiller, 2nd ed., 2005, p. xviii (Princeton University Press, 2005).

across many aspects of the global oil industry and the globalised financial markets, but we can make a modest start here and now. For if a picture tells a thousand words, our chart here also furnishes immediate and very eloquent support for the bubble theory.

The shaded band below the actual peak of the oil price in Figure 1 sketches the matching shape of one of our more notorious speculative bubbles of recent years: the Nasdaq tech stock/dotcom share boom in the US which peaked in March 2000 before its own spectacular bust.

Perhaps no two investment propositions could be more different than crude oil and the boom-era Nasdaq. One is the physical product of a classic, grimy, "old economy" business focused around costly and dangerous real world engineering projects in often inhospitable environments; the other was a basket of dotcom internet retailing, biotech, new media and other similarly "weightless" members of a supposed "new economy". Yet as Figure 1 indicates, both the oil price and the Nasdaq 100 shared a strikingly similar trajectory through the crucial period spanning their most rapid appreciation and subsequent abrupt collapse.

This is more than just a matter of resemblance from a distance. Figure 2 details this uncanny coincidence in tighter focus (and without the Nasdaq vertical axis offset used in Figure 1 for ease of presentation). It clearly shows how closely the two markets shadow each other as they move through their respective zeniths, years apart.

Lest those rightly wise to the tricks of statistical presentation begin to question this startling parallel, it should be stressed – no axis-shifting or otherwise underhanded data transformation is required to draw this graph. It is simply a fact that, although separated across time, space and underlying investment assets, both the Nasdaq 100 and the oil price gained, gained again more rapidly, and then dropped precipitately through near-enough the same percentages in three distinct phases over the same number of weeks in their respective boom to bust cycles. We shall examine an extended version of this chart later, but what can better explain this coincidence other than some unfolding logic of speculative bubbles, an underlying algebra of investor herding common to all such episodes? The spectacular rise and fall of the oil price certainly looks like a speculative bubble, if recent history is any guide.

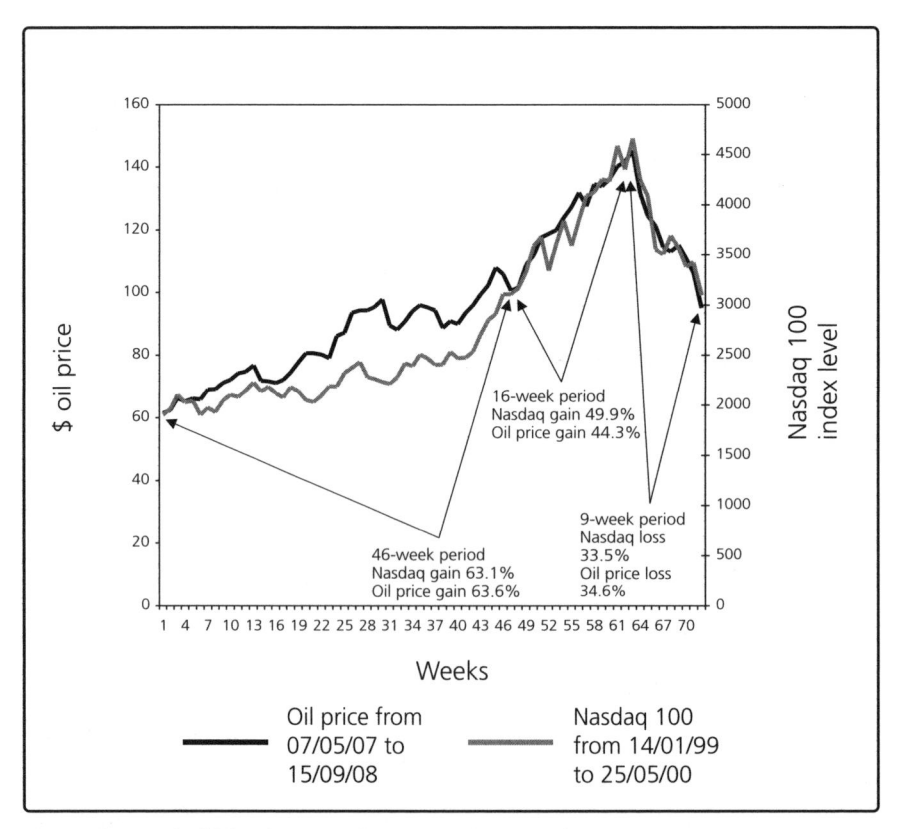

Figure 2: Two bubbles [Source: Thomson Datastream]

Does it matter if the oil blow-out was indeed a genuine speculative bubble, a fragile 'naturally-occurring Ponzi scheme' to use Robert Shiller's terminology, demonstrating dynamics of the type described variously both by Shiller and the late Hyman Minsky, another major theorist of speculative excess? That is, was it merely a self-referential feedback loop of ever-higher valuations, only ever justified on the basis of the next herd of investors buying into the latest "this time it's different" story? And if so, so what? Who cares that some made fortunes while others lost their shirts in the sort of investment craze which is, after all, now widely recognised as an all-too-periodic visitation upon our market-based variant of economic development? So what if the petromania of 2008 now takes its place in the historical parade of previous investment manias, including such celebrated episodes as the Dutch tulipomania of the 1600s, the "Mississippi Scheme" of early 1700s France, and the roughly coincident episode in England – which bequeathed us the usage by which we now tend to identify similar market phenomena – the "South Sea Bubble"?

Black Gold, Devil's Excrement

It matters greatly, as the list of charges already laid against the high oil price underlines. No other commodity can affect the world economy the way oil can – certainly not copper, gold or platinum, other important commodities which also experienced astounding price blow-outs through roughly the same period. Yet through 2008 they were not the topics of conversation around the water cooler, or on radio talk shows, or TV news reports, in legislative hearings held by elected politicians, or at inter-governmental summits. Whereas oil most definitely was centre stage across all. The reason is simple. At the dawn of the 21st century, over a hundred years since combustion engines first shattered the bucolic peace of our forefathers, the old-fashioned, literally prehistoric, black, slick, smelly liquid hydrocarbon fossil fuel known as crude oil remains our industrial lifeblood, and therefore the single most-traded commodity worldwide both in terms of volume and value.

Fuel products derived from crude oil drive our transport on land, sea and in the air – while also firing a good portion worldwide of our electrical power generation capacity. Meanwhile materials derived from crude oil and its associated natural gas both underpin the ubiquitous use of plastic throughout our industrialised society, and also provide us with a plethora of synthetic fibres to clothe us in evermore imaginative materials. Ever since the then-dominant British naval fleet made the momentous decision to switch its boilers from coal-fired to oil-fired at the beginning of the 20th century, access to and control of supplies of crude oil has determined the weightiest of foreign policy decisions made by states, and decided the outcome of wars fought between them.

The incredible wealth oil can bring to those who control it has funded the rise of fabulously rich dynasties around the world, from the original US oil barons of the early 20th century to the Gulf sheikdoms of the contemporary Middle East. However, these riches also mean corruption, crime and violence are frequently inextricably intertwined with the production and exploitation of oil. It is now common to refer to the "resource curse" afflicting developing countries such as Nigeria, endowed with significant oil resources but lacking the requisite governance structures to check the potential for embezzlement, bribery, thievery and outright armed conflict, both inter-state and civil, that the promise of such resources can awaken in people. Oil is indeed commonly known as "black gold", but it has also famously been described as 'the devil's excrement' – by a one-time president of oil producer cartel OPEC, no less (former Venezuelan oil minister Juan Pablo Perez Alfonso, speaking in 1975).

A high oil price enriches some countries, but also spells relatively higher costs across practically the whole gamut of business activity worldwide. It brings an increased risk both of knock-on inflation in the economy, yet also recession due to there being less cash to spare in oil-importing countries for other goods and services. Together these two outcomes spell "stagflation", the term coined in the 1970s following the "oil shock", an embargo by Arabian oil producers in protest at Western support for Israel that signalled a new heavyweight status on the global stage for OPEC. Much as it might pain us to admit it, nothing matters as much to the smooth functioning of our society as our supply of crude oil, and no price is therefore more important to us than the price of crude oil.

Follow the Money

Whoever can influence the oil price wields enormous power over the direction of our globalised economy – which is exactly what this book is about. Yet this is a tale played out a million miles from the swamps of the Niger Delta, or the snows of the Russian Far North, or the deep water offshore of Brazil, or the heat of the Arabian Desert. This is a tale played out on computer screens across the world, shouted hoarse across City trading pits, and buried in the annual statement of returns from your own pension fund. For the truth is that, as with so many other spheres of modern life, the globalised financial investment sector has effectively colonised the oil market as a prime source of speculative return.

As a result, many people are worried that the price of this most crucial commodity of all now dances to the tune of investment manager sentiment, herded in and out of short-lived trends, subjugated to trading expectations regarding comparative yields across asset classes that also include equities, bonds, currencies, precious metals and other commodities. In the worst-case scenario, this "financialisation" of the oil market causes the price to lose touch with the actual underlying physical fundamentals of supply and demand. And certainly, this is what the evidence collected here suggests happened over summer 2008.

For many, it seems economically suboptimal – not to say morally unfair – for the self-serving investment strategies of a well-paid transnational financial elite to dictate wild gyrations in the oil price. It rides roughshod over the common interest that the vast majority of people worldwide share in enjoying relatively stable energy costs. Yet this is the natural outcome of the particular way we have organised our "market-capitalist" economy to date – and ultimately, changing this state of affairs is a political question. It was one which many politicians themselves became very interested in as fuel prices rocketed skywards through early 2008. In the US in particular, public opinion on the matter forced a series of legislative hearings that sat across many months in both Houses of the US Congress. The testimony offered and evidence submitted to these hearings loom large in this tale, as do the responses engendered from US market regulators.

Notwithstanding this flurry of legislative interest, however, one year on there has yet to be any concrete action taken to control the financial sector influence on the oil price which drove this chain of events. This is also despite the previously avowed intent of new US president Barack Obama to rein in such speculation in the oil market, a pledge he made when campaigning for office. In mid-June 2009, the White House unveiled a package of proposed regulatory reforms for the financial sector, intended to ensure that the widespread institutional failings that caused the global credit crunch, its attendant banking crisis and the economic catastrophe we are now enduring can never happen again. Re-regulating financial sector interest in the commodity markets, however, is markedly absent from these proposals. Even though White House spokesman Robert Gibbs told reporters, 'I don't think the President's concern has changed,' he admitted he did not know if speculation in oil markets would be dealt with in further proposals.

It is easy to see why such steps have slid down the presidential "to do" list. The priority for the world's most powerful man right now is stabilising the global financial markets at large. And while commodities such as oil are an important part of this panorama, attention is fixed instead on a whole other vista of market failure, the delinquent peddling of sub-prime mortgages. The resulting toxic asset infection was spread by the virus of securitisation and similar credit derivatives throughout the length, depth and breadth of the banking sector worldwide, and remains as yet largely unresolved. Significant as it is, the dramatic oil price spike of summer 2008 risks becoming the other, forgotten financial excess of the late noughties – overshadowed by the all-encompassing fallout from that more obvious bubble of our times, the worldwide real estate boom-and-bust.

That would be unfortunate, as the lessons we can derive from studying the petromania of 2008 are needed right here, right now. Incredible as it may seem so soon after the oil price collapse in late summer 2008, the financial forces which blew that bubble are at work once again, stoking the next flare-up in oil, even as physical market fundamentals look at least as unsupportive as they did last year, and possibly more so. Oil doubled from its Christmas 2008 low, to around $70 per barrel in mid-2009, and investment bank cheerleaders are once again pencilling-in forward estimates of $95 per barrel for mid-2010. Triple digit price forecasts will probably be in vogue once more by the time this volume has gone to print. Back in the real world, however, politicians and economists warn that this unchecked and patently illogical price appreciation is undermining recovery prospects across the wider economy. Following its initial appearance in summer 2008, petromania may well come to be characterised by recurrent outbreaks of the fever. Understanding how it progresses is at least half the battle in tackling it.

1

The Ascent

'At first, as in all these gambling mania, confidence was at its height, and every body gained.'

Charles Mackay, "The Tulipomania" in *Extraordinary Popular Delusions and the Madness of Crowds* (1841)

1.1 The Triple-Digit Threshold

Global markets opened 2008 in a jittery mood as January 2, the first trading day of the New Year, saw leading share indices stumble in turn across the time zones. While the main equity boards in Hong Kong, Frankfurt, Paris and London would all close down more or less a percentage point, the real upset was in New York where, following the release of particularly weak US manufacturing data, the Dow Jones Industrial Average was heading for a 1.7% drop – its worst first-day percentage performance since 1983. Yet elsewhere in the city, another market was about to breach a more momentous threshold. Shortly after noon, a commodity broker working on the floor of the New York Mercantile Exchange (NYMEX) became the first person in history to pay $100 for a barrel of crude oil.

Stephen Schork, a former longtime Nymex trader who now edits the daily energy market analytical newsletter *The Schork Report*, remembers the exact moment: 'I was on the telephone with a journalist and I think the bid in the market was around $99.30 or $99.40, so he was asking me "When do you think we will see $100?" And as soon as he asked that, a $100 print went on the board and I said, "Yep, right now I think we'll see it…"'

The trade immediately made headlines worldwide. As the reporter's interest shows, crude oil's rapid appreciation, from around $60 per barrel a year earlier, was already big news, and that first hundred dollar deal on Nymex – the exchange the wider world looks to for its benchmark oil price – was immediately controversial. Not just because the new era of triple-digit oil prices it heralded was feared both by ordinary citizens struggling with sky-high fuel costs, as well as economists and governments concerned at how appreciation in such a ubiquitous industrial cost input was stoking global inflation. More immediately, there were questions regarding the legitimacy of the transaction and whether it should stand in the trading record.

Those physically present in the Nymex exchange could see the agreed $100 price printed up on the board over the open-outcry "bear pit", where the floor traders shout bids and offers at each other. However the bulk of crude oil trade settled on Nymex is not transacted on the exchange floor but through Globex, an electronic trading system that links traders all over the world into the same market through their computer screens. Nymex floor and electronic trading are ultimately settled as a single market, and barrels bought or sold vocally on the exchange floor are fungible with those bought or sold electronically off-floor. But technically, on-floor and off-floor are two separate pools of liquidity, so prices on Globex may not exactly match those posted in the exchange. This is what happened with the first ever $100 barrel: most market players, following trading at their screens elsewhere in the US or abroad, simply never saw it.

Yet the deal looked distinctly anomalous even on the floor of the exchange, where transactions both immediately before and after it were at significantly lower prices, around the $99.30-$99.60 per barrel range. This was an "out-of-market" trade questionable enough for Nymex to investigate – but it was confirmed in due course as valid.

Nevertheless, the suspicion was that the deal had been done simply to grab the "bragging rights" on the first ever $100 oil trade. Within hours, unnamed sources were quoted in news reports to the effect that the oil had been bought by a "local", a floor trader who operates using his own rather than client money, that he had bought the minimum permitted lot of a single contract for 1000 barrels from another local, and that he had turned around and sold out of the position immediately afterwards. Mainstream media, including the *Financial Times*, were happy to pick up Schork's own comment that same day: 'A local trader just spent about $600 in a trading loss to buy the right to tell his grandchildren he was the one who did it. Probably he is framing right now the print reflecting the trade.'

The local supposedly responsible for this stunt was quickly named as Richard Arens. Arens is indeed a Nymex floor trader, but has himself never publicly confirmed nor denied his role in the episode. Yet that reticence did not save him from a tsunami of public scorn for his supposed "vanity trade". 'We're sure the girls at the bar will be real impressed,' was the comment on the *New York* magazine website a day later, under the headline 'Richard Arens is Having His Moment'. Indeed, Arens briefly became something of an online hate figure, with US bloggers calling for his face to be put on dartboards or even for the man himself to be burned at the stake – if not with anything as expensive as gasoline. Outraged at the idea of one man pushing up already crippling fuel prices for his own selfish ends, this blogger slant played well with the theme of financial whiz-kids wreaking havoc in the lives of ordinary people, so resonant since the onset of the banking crisis the previous August and the first intimations of the current global economic turmoil.

Yet what if Arens was unfairly pilloried? Not so much in the sense of mistaken facts – although we should note that an alternative tale emerged in the *New York Post* days later, with Arens as the seller, and the buyer instead a large commercial entity. Rather, in the sense that even if Arens was indeed the $100 barrel buyer, could he not just have been doing something quite normal for a trader – trying to get a feel for what the market really wants? After all, the oil price had been extremely volatile since punching through $90 per barrel the previous autumn. While it had not yet traded at $100 per barrel, it had come very close a couple of times already, trading as high as $99.29 in late November, and bouncing around the $95-98 range in the last few trading days of December. If Richard Arens was indeed the man who did the deal, we may never know his motivation as, well over a year after the event, he still refuses interview (I was informed of this through the intermediary of a Nymex official). Yet we can entertain definite ideas about how the trade might have occurred.

Firstly, as said, a market participant may have just offered the bid to get a feel for the market – this is how a true open outcry auction market works. Imagine that someone, wanting to know how the commodity in question is trading, calls down onto the trading floor, 'Where's the market?' The answer comes back, '30 bid, at 34.' In other words if you want to sell, someone will buy from you at $99.30, and if you want to buy, someone will sell to you at $99.34. This is an important point – a deal can be done on the basis either of someone having made a bid to buy, or someone having made an offer to sell the asset being traded. But there is no requirement that any trader both bid and offer at the same time, and it is possible to offer on the heels of someone else's bid: the hallowed, age-old conventions of open outcry trading stipulate only that both the best bid and the best offer at any one time silence all others. As Stephen Schork recalls, 'Given the volatility, I don't think anyone was willing to sell at that point – so the market was only being bid, at around $99.30. And then someone probably just said "At 100," and this Arens guy probably just said, "Buy it!" He would not have bid $100. I'm guessing someone just threw out a $100 offer, and this guy took it. Maybe he was doing it because he felt there were a lot of stops there, and this would create a rally.'

Stop orders are orders from clients that brokers must act on as soon as the market price moves through a certain price level or "stop". If many clients have placed orders to buy or sell at a particular price threshold, then once a transaction has been realised at that price in open trade, all brokers with stop orders at that level are technically obliged to complete the order as soon as they can, regardless of the fact that a wave of large buy orders emerging once a particular stop is hit can lead to the price quickly exceeding the stop level. This is a basic rule of markets we shall have cause to return to in due course. Of course, by definition, for every buyer there must be a willing seller; but if a relatively large number of buy orders emerge at any one time, lots already offered at a certain price are quickly snapped up, and the next wave of sellers to emerge in response to the demand will of course pitch their bids higher again, and so on until the buying runs out of momentum. So it would have been solid trader logic to take up the $100 offer when a subdued sell-side in the bear pit meant that, for however brief an interval, this was indeed the best offer in the trading ring.

Traders know full well that "taking out" stops can lead to rallies, and as far as Schork is concerned that was what the buyer who gave the nod to 'at 100' was trying. 'It's analogous to a poker game – you call a certain raise because you want to test the structure of the other guy's hand, the other players. You bid, you offer, and you bid; because you're pushing against the market, you want to see where the market's weak and then you'll push. So there's nothing wrong with what he was doing.' Oddly enough, despite the obvious psychological significance of the triple-digit price level, on that particular day there was not a wave of "stopped-out" buying when the $100 trade was hit inside the exchange. And this, too, also explains why the trader in question would indeed probably have sold out of the position as soon as possible afterwards. As Schork says, 'The market immediately traded lower, but that's what happens. You think, uh-oh, I made a mistake and you get out – I'd rather lose $600 than $6000 or $60,000. So he tried it, he took a $600 bet and it didn't pay off.'

Despite the media mood within which his comments on the actual day of the trade were quoted, Schork in fact empathises with whoever ushered in the era of $100 oil. 'I don't think he ever envisioned the snake pit that he opened up when he did this. He was doing something very normal for what goes down on the floor – probing the market.' And while the mystery trader – whether Arens or not – faced derision for gratuitously trying to push oil prices higher, in reality he was only guilty of bad timing. Oil actually closed at $99.62 on January 2 2008, but the very next day it once again breached $100, trading at $100.09 intraday. Yet again, the notional wall of stops was not triggered and the price once more slid back, closing at $99.18. After these damp squibs, there was a pullback in prices to around the $90 mark for the rest of January. But February 19 2008 saw Nymex oil finally log a closing price above $100, at $100.01 – and from there, as Schork says, 'We were off to the races.'

Thus far, oil had still not reached an all-time price high in inflation-adjusted terms. The price spike in April 1980, as revolutionary turmoil gripped key oil-producing state Iran, was deemed to have reached the equivalent of around $103 in current money (although there are varying figures given in this regard, depending on how inflation is factored in). Whatever its exact inflation-adjusted level, that 1980 record soon tumbled as oil breached $110 per barrel in early April 2008 and $120, a month later, in early May. By July 11 2008 the price had gone as high as $147.27, and a break through the $150 mark was seen as inevitable. It was hard to remember that oil had cost less than half that just a year earlier – but by then, of course, there had been a lot of Nymex crude oil trading tickets printed at $100-plus prices, and the novelty value had worn off.

1.2 Black Gold, Diabolical Prices

By the beginning of 2008, the oil price had been worrying a lot of people for quite some time. Throughout the 1990s, the price of oil had consistently traded around an average of $20 per barrel. But as the 21st century began to unfold, crude started to probe higher. It was averaging around $32 per barrel in 2003, then $42 in 2004, before jumping to an average of $57 per barrel in 2005. In 2006 the price of crude had averaged $66 per barrel; and in 2007 this average had climbed again to $73 – an unthinkable level just a few years previously. And as we know, it was set to break fresh records in 2008. What on earth was driving such rapid price appreciation in the world's most important commodity? A complex of reasons, relating to the fundamentals of physical supply and demand for oil, was widely seen as underlying this phenomenon.

Strong economic growth worldwide

The period from around 2002 onwards had seen strong economic growth across the globe – crucially not just in advanced countries but also in the developing world, as mass industrialisation and urbanisation finally took off in both China and India. Oil itself is of course key to economic development in terms of industrialisation and urbanisation, and economic growth of this kind spells growth in oil consumption.

Using broad-brush, yearly average figures from BP's *Statistical Review of World Energy*, in 1997 total world oil demand was around 73.6 million barrels per day (mbpd). By 2002 this had grown to 77.8mbpd, a 6% increase over five years. Through the next five years to 2007, however, this total world demand had grown to 85.2mbpd, a 9.5% increase.

China's inexorable economic rise was particularly noteworthy in this regard. From being the sixth-largest economy worldwide in 2004, just two years later in 2006 its economy was deemed to have displaced both France and the UK to become the fourth-largest, behind the US, Japan and Germany. Historically self-sufficient in its oil supplies until 1993, by the beginning of 2008 China was seen to have surpassed Japan to become the second-largest oil importer in the world, after the US. The same BP dataset as quoted above estimates Chinese crude oil demand at around 4.2mbpd in 1997. Five years later, in 2002, it had grown to 5.3mbpd, a 26% increase. By 2007, it had reached 7.9mbpd, a 49% increase in five years. Or, to look at the hungry dragon's appetite from another angle (and perhaps that most often quoted by newspapers), from 2000 to 2007, the Chinese crude oil demand growth of 3.1mbpd was equivalent to 39% of total world crude oil demand growth in the same period.

All else being equal, for oil prices to remain constant the rising demand due to increased economic growth must be met with sufficiently increased supply from production. The US government Energy Information Administration (EIA) is another major and widely-quoted source for energy market statistics which, in contrast to BP's annual review, presents its data on a quarterly basis. As might be expected with such a real-world data gathering exercise, there are discrepancies between the two sets of figures. Yet both agree that in 2006 and 2007, at least, world oil demand averaged more than world oil production. BP figures give 2006 and 2007 production of 81.7mbpd and 81.5mbpd respectively, versus demand of 84.2mbpd and 85.2mbpd respectively. The EIA figures are production of 84.5mbpd and 84.4mbpd respectively, versus demand of 85mbpd and 85.9mbpd respectively. For what it is worth, a third set of figures from another reputable data source, the OECD-sponsored International Energy Agency (IEA), sees 2006 production of 85.5mbpd more than meeting demand of 85.1mbpd, but 2007 production of 85.5mbpd again falling short of demand at 86mbpd.

What do these figures tell us, apart from the important lesson that there are multiple versions of even the most basic truths regarding oil supply and demand? They support the view that from the mid-noughties onwards, growth in oil production was apparently not able to keep up with growth in oil demand, particularly in light of the trend that Chinese demand appeared to be on. Whatever excess demand there was over production through these years would of necessity have been met by a net drawdown on inventories of oil in storage. Of course, as their primary supply and demand figures differ from year to year, so too do these implied stock drawdown figures from BP, the EIA and the IEA. But regardless of the exact figures, year after year of net drawdown on inventories is not a sustainable situation, as this stock cushion will eventually be exhausted. Here, then, seem solid grounds for rising oil prices, as increasing demand chases scarcer supply in a context of diminishing inventories – but what were the reasons for the perceived global production shortfall in the first place?

Tightening OPEC supply and capacity

The acronym OPEC – standing for the Organization of the Petroleum Exporting Countries, the oil production cartel dominated by Arabian petro-states since its founding in 1960 – is synonymous with control of the world's oil supply and the pricing thereof. The fact of the matter is that OPEC countries (currently comprising Algeria, Angola, Ecuador, Iran, Iraq, Kuwait, Libya, Nigeria, Qatar, Saudi Arabia, the United Arab Emirates and Venezuela) ostensibly control as a unitary entity the total amount of oil they produce as a group, in the interests of managing supply in relation to demand so as to maintain prices at levels that offer these countries a reasonable rate of return on their oil sector investments. OPEC can do this because between them its member countries produce somewhat over 40% of global oil supply; and counted as a cartel, OPEC is the largest single contributor to that global oil supply.

Despite the general upward trend in oil prices since 2000, late 2006 actually saw a significant pullback in oil prices, from around $74 per barrel in late summer to just over $60 by the end of the year. OPEC member countries had become accustomed to the $70 per barrel range, and many were planning their budgets around this price level, so the slide discomfited them enough for the cartel to announce production cutbacks in November 2006 totalling 1.2mbpd. By February 2007 the oil price had actually dipped below $60 again for a spell, so OPEC announced further cutbacks totalling 0.5mbpd. Actual compliance by OPEC member countries with cartel quota orders is variable at the best of times, with overproduction beyond quota anecdotally endemic. Such cuts also take some time to feed through into actual production, even given willingness to comply. Nevertheless, for many industry observers it was these OPEC cuts that played the biggest part in a perceived tightening of the global oil supply and demand balance from late 2006 into 2007.

In one sense, this should not have been too much cause for concern. As throughout much of its history the task of maintaining oil prices at what it saw as reasonable levels involved OPEC instructing its member countries to produce less – sometimes considerably less – than they were actually capable of doing, the cartel is also traditionally seen as the "swing" supplier in the global oil market, the player with the ability to easily bring more barrels to the market should these suddenly be required. Oil importing nations in the developed world have become used to lobbying OPEC vociferously to increase its supply quotas whenever oil prices seem to get too out of hand, and despite popular Western resentment of the oil sheikhs supposedly driving around in their gold-plated Rolls-Royces at our expense, OPEC has in general proved itself a reliable partner in easing the market when asked to do so by developed world economies.

It could hardly be otherwise when the United States, the number one oil consumer and importer in the world, is the military protector and sponsor of the Al Saud dynasty that rules Saudi Arabia, the number one oil producer in OPEC and the member country that dominates its affairs by dint of this preponderance (according to the EIA, Saudi has generally been producing around 8-9 million barrels of crude per day in recent years, more than twice its nearest competitor in the group, Iran, on some 3-4 million barrels per day). OPEC is also seen as being sensitive to the threat of permanently high prices increasing both marginal supply viability and fuel substitution. In other words, if the cartel maintains prices at too high a level it runs the risk of these high prices encouraging new oil exploration and production elsewhere in the world, which could in due course take market share from its own member countries. It might also risk encouraging consuming countries to switch an increasing portion of their own energy supply to non-oil sources, again harming its own long-term prospects for market share. Such substitution can involve renewable and sustainable technologies, including the crop-derived ethanol that has become such big business in the US and Brazil in particular.

Beyond this there is of course the fundamental issue that when oil prices rise too high, they can choke off economic activity in absolute terms, and potentially thereby help cause a recession – a situation which is never good for OPEC revenues. In short, then, OPEC is well aware of the imperative not to kill the goose that lays the golden egg. So on the face of it, if OPEC production cuts were driving oil prices to unbearably high levels for the global economy, all that had to be done was convince OPEC to bring back on stream some of the spare production capacity it had sitting idle as a result of prevailing production restrictions. But what if OPEC itself did not really have so much spare capacity available after all? If its own announced cuts feeding through into the global oil supply and demand balance were one reason given for oil prices trending higher, doubts that OPEC could actually add that much to global supply, even if it reversed those cuts, was another reason touted by many for this rapid oil price appreciation.

As crude oil demand grew steadily year-on-year, and actual OPEC production was forced to gradually grow in tandem, it turned out that what had originally been a significant quantity of surplus production capacity, above quota and spread across the cartel (of perhaps some 10mbpd through the 1990s), had shrunk considerably. So much so that what spare capacity remained was largely concentrated in just Saudi Arabia. According to the EIA, in 2002 the cartel was operating on average with over 5mbpd of spare capacity, equating to some 6% of the then-prevailing global demand of around 77mbpd. By 2006, however, surplus supply capacity available to the global market from OPEC had shrunk to below 2mbpd – and indeed at this point many commentators felt OPEC had "lost control" of the oil price through having allowed its swing production potential to drop to just over 2% of then global demand, running at around 84mbpd. By the end of 2007 the EIA reckoned OPEC spare capacity had recovered slightly, to somewhere in a range between 2-3mbpd – with practically all of this being in Saudi Arabia. Saudi capacity in late 2007 was stated to be around 11.3mbpd and planned to grow to 12.5mbpd in 2009. At the same time, the EIA estimated actual Saudi production running at around 8.6mbpd.

Disappointing non-OPEC production

Notable volumes of crude oil production outside the OPEC umbrella come from other nations and regions. Indeed, according to IEA figures which strip out natural gas liquids (NGL) production, despite the post-Soviet decline of its hydrocarbon industry Russia is arguably the largest *actual* producer of crude oil worldwide – as opposed to Saudi Arabia, which has both the largest *potential* crude production capacity plus significant *actual* volumes of NGL production, which some other market-watchers (BP, for example) simply lump together with crude to label the country the largest producer, full-stop. From 2005 onwards, however, Russia has consistently produced close to or just over 10mbpd of crude oil, depending on whose figures you take.

Likewise, the Caspian region of former Soviet satellites such as Azerbaijan and Kazakhstan, the North Sea, the Gulf of Guinea basin and the wider West African coastal zone, as well as the Gulf of Mexico offshore of the US, are all important sources. It is a standing structural presumption of oil market analysis that, in contrast to OPEC members, such non-OPEC producers are always working at full capacity – in other words, they are always producing as much as they can at any given moment. The logic behind this notion is that if the market-dominating OPEC is in any case always going to adjust its own production to maintain prices at its preferred level, then non-OPEC countries have no reason not to pump at capacity and just take whatever price is on offer at any given time.

Despite this price-taker incentive for maximising non-OPEC production, total non-OPEC production has disappointed expectations through recent years. In several key producing regions, such as Mexico, or the UK and Norwegian North Sea zones, output rates at ageing fields are well into their natural decline as the resource in question nears depletion. Indeed the steep drop-off in Mexican production has surprised experts. Meanwhile, numerous new large projects in the various non-OPEC production hotspots have been bedevilled with various development problems and delayed start-ups (for example, hurricanes damaging in-construction platforms in the Gulf of Mexico). For 2008, non-OPEC production was still expected to rise incrementally by many analysts – the EIA foresaw growth in non-OPEC supply of around 0.9mbpd, this coming primarily from projects in Brazil, the US, Russia and Canada. Yet after a string of surprises on the downside, the feeling among many analysts was that not much faith should be put in non-OPEC supply projections that presumed the smooth development of ongoing projects in accordance with their stated timetable.

The shifting dynamics of non-OPEC production also gave grounds for expectations of higher oil prices through arguments relating to the "marginal cost of supply". This is the idea that oil should naturally be priced at the production cost of the "marginal" barrel, the production cost of the last barrel required in building the supply stack sufficient to meet global demand. Such supply stacks begin at the bottom with the easiest, lowest-cost production, which yields a good return even at relatively depressed oil prices – some production in the Middle East can be costed in single-digit dollars per barrel. As demand moves beyond the limits of such capacity, supply has to be sourced from evermore difficult and costly production frontiers – from drilling at greater depths beneath the terrestrial crust into technically more challenging high-pressure, high-temperature fields; or drilling locations deeper or further offshore, or in more remote and inhospitable terrains such as the Arctic north. As depletion of existing resources leads companies to source more and more of their production from such terrains, the marginal cost argument dictates that the price of oil will have to rise sufficiently to reward the higher costs incurred in exploring and producing these new resources.

In general, oil in the core OPEC region of the Middle East is easy and cheap to produce – one reason why the Mideast oilfields have been developed so prolifically, and have been in production for such a long time – while the more challenging production environments tend to fall into the non-OPEC categorisation (although this is not an absolute distinction). As noted above, at the start of 2008 the EIA was still expecting non-OPEC production to achieve some minimum level of net growth, and a good chunk of this was seen as set to come from Canada. However, this Canadian supply is not from "conventional" oil drilling, but comes instead from a non-traditional, "unconventional" (to use the specific oil industry jargon) supply source. These are the so-called tar sands or oil sands deposits, in which oil is locked in an extremely viscous, bituminous form that demands a lot more engineering than conventional oil in extraction, as well as higher operating and processing costs. By the start of 2008, an oil price sufficient to reward increasing production from Canadian oil sands was seen to be around the $70 per barrel mark at least, and due to rampant construction cost inflation in the Alberta region, where the industry is centred, some new projects needed closer to $90 per barrel to ensure economic viability.

Some analysts are scathing of such marginal cost arguments. 'The marginal barrel is always in the Middle East,' is a typical rejoinder. Meaning, in other words, that all such arguments are rendered moot by the fact that all it takes is for OPEC to increase its production quota at any given moment, and this fresh OPEC production will immediately come into the supply stack at the lower end of the so-called "cost curve" of producers; thereby removing the need for an equivalent amount of production at the top of the curve, and by extension any need for prices to match the costs of that displaced production. However, as we have seen above, doubts have emerged over the ability of OPEC itself to continue to push and pull production in and out of the supply stack at will and, as we shall see below, some even go further, and say this ability is already illusory. In a world where OPEC is already pumping at or near full capacity, the marginal cost of supply argument does start to make more sense with regard to oil price levels.

Geopolitical conflict

By early 2008, war and the threat of violence had hung over the oil markets for quite some time. Of course, the US-led invasion of Iraq in 2003 had itself reduced production in that OPEC member state, and to even more negligible quantities than the Saddam Hussein regime had managed under a decade of sanctions. Elsewhere, a serious insurgency on the part of local people against international oil company interests had flared up in the Delta region of Nigeria. A militant group called MEND (the Movement for the Emancipation of the Niger Delta) had gradually emerged at the forefront of a campaign of sabotage and violent gun and bomb attacks. This campaign was now choking off production from this key OPEC contributor – as facilities and pipelines were damaged, and oil companies were forced to evacuate staff and shut down production due to the security situation. By early 2008 it was reckoned that the security situation in the Delta region had forced the suspension of some 500,000 barrels per day from the typical Nigerian capacity of around 2.5 million – in other words, a fifth of potential production was interdicted by the violence.

Then there was Iran. The Shia Islamic republic was by then well into its still-ongoing face-off against the US and its regional ally Israel over the alleged Iranian programme to develop a nuclear weapon. Sabre-rattling by either side periodically ratcheted up the tension and led analysts to wonder what a US or Israeli attack on Iran (as noted above, the second-largest oil producer in OPEC) would mean for the global supply and demand balance, and likewise how many oil tankers cruising through the Persian Gulf the Iranians might be able to sink in retaliation. From at least 2006, the ascent of the crude oil price had played out against a chorus of US-Iran war predictions – such that, as early as January 2007, one analyst was telling this writer personally that it was only the perceived risk of war being launched against Iran that was supporting oil prices (then around $60). One year on, oil was a lot higher, Iran was still defying the US over its nuclear intentions, and in response the world's most powerful state was still 'keeping all options on the table', as the phrase went.

Meanwhile, the "War on Terror" launched by the US against radical Islamist armed groups, in the wake of the 9/11 attacks in 2001, had entangled it in both Afghanistan and Iraq, and invited retaliation against US and allied interests around the world. In this context, the US dependence on imported oil was seen as a major vulnerability – and explicitly highlighted as such by America's enemies, with Al-Qaeda leader Osama bin Laden, himself a Saudi, famously listing oil refineries as among the 'hinges' of the world economy his disciples should target. Accordingly, from 2003 onwards in Saudi Arabia there was indeed a string of attacks by Islamist terrorists against oil industry facilities, with the intent of destabilising the all-important Saudi supply. These had not yet made a serious impact, but the potential threat was still perceived as very serious. In Central Asia and the Caspian region too, the US was on the alert against possible attacks against oil and gas industry infrastructure, and took steps to mitigate threats – for example, by funding the battalion of commandos specially set up by the government of Georgia to guard the 1mbpd-capacity Baku-Tbilisi-Ceyhan (BTC) pipeline, as it shipped oil across its territory toward Western markets.

1.3 Grand Narratives

A background was therefore clearly in place by early 2008 against which the strong appreciation of the crude oil price could be justified. The key elements in this picture were strong global economic growth led by emerging Asian countries and the perceived inability of global supply to keep pace with resultant demand, the latter due to a combination of tightening OPEC supply and capacity, disappointing non-OPEC supply performance, and the heightened possibility of violent disruptions to actual daily supply. Yet on the bedrock of these fundamental observations were also constructed various analytical superstructures, a number of sweeping grand narratives which informed various shades of opinion as to how the oil markets were developing given this agreed background, and therefore how the oil price would move in the coming years. We can summarise these narratives as follows.

Emerging Asia and the "commodity super-cycle"

Inevitably, societies undergoing the twin transformation of industrialisation and urbanisation experience an increased intensity of commodity usage inherent in building all those factories, skyscrapers, power plants, railways, roads and houses. This means persistent high levels of raw material demand being maintained over a long period (the postwar transformation of Japan is often cited as the classic case in point). Such a period can be described as a "commodity super-cycle", an idea pushed by analysts who insisted the sky-high prices seen in recent years for such essential industrial inputs as oil and copper would persist indefinitely, since industrialisation and urbanisation in emerging Asian superpowers such as China and India marked an irreversible, structural boost in demand. This demand boost or "super-cycle" would supposedly persist for decades, regardless of the vagaries of the developed world's business cycle, and its influence on global commodity supply and demand balances would maintain prices at a considerably higher plateau – higher than those seen in recent years – for the foreseeable future.

With regard to oil, the commodity super-cycle contains a particular twist beyond just the simple first-order hunger for crude oil in general. It is peculiar to emerging Asian growth, and China in particular, that diesel is the refined fuel of choice. (This is similar to continental Europe but in contrast to the preference for gasoline in the US.) This means that in considering the effects of Chinese demand on crude oil supply, special attention is paid to the section of that crude oil supply which is most easily refined into diesel. In practice, this means a relative preference for so-called "light, sweet" crude as opposed to "heavy, sour" crude. Oil that is "heavy", relatively dense or viscous, and/or "sour", relatively high in sulphur content, is more difficult and costly to refine than "lighter" and/or "sweeter" crude. It is more difficult to refine both diesel or gasoline alike from heavier, sourer crude oils, but while there has historically nevertheless been significant investment in refining capacity capable of "cracking" gasoline from heavier, sourer crudes, there is a relative lack of capacity configured to refine diesel from such oil. Chinese hunger for diesel was therefore seen as equating to a squeeze on the light, sweet end of crude oil production in particular.

Peak oil

It goes without saying that as there is a finite amount of crude oil trapped in geological formations around the world, it is theoretically possible to use it all up one day – but that day remains a long-distant reckoning, even now. The important question is actually not when will the geological supply of oil run out in absolute last-drop terms, but instead when and at which level will global oil production reach what turns out to be its absolute historical peak in daily output. After all, it is this peak level of daily global supply, and how long it persists for, which will limit the amount of crude oil-dependent industrial activity that our global society can support at its current technological level of development. When this limit is reached, not only will additional growth in oil-dependent activity become a zero-sum game, where one consuming country's gain will necessarily be another's loss – but, according to some, the statistical history of oil production patterns indicates that actual decline in production levels follows on inevitably from this peak with grim predictability. Peak oil theorists argue that we are already close to or even past the point of maximum possible global oil production, and therefore heading over the hump and into the downward-facing slope of "Hubbert's Peak".

Hubbert's peak, also referred to as his "curve", is a bell-shaped statistical distribution seen to describe the characteristic profile of daily production over time from an oilfield and, in aggregate, from many oilfields, from a wider oil-producing region, a whole country, or even the world. It is named for the US petroleum geologist M. King Hubbert, who used this model to correctly predict more than a decade in advance the point in the early 1970s at which oil production across the lower 48 US states actually reached its peak, before decline set in. Peak oil advocates apply the same modelling techniques (using variations on the several bell curve distribution shapes known to statisticians) to known global oil reserves and their dates of discovery. This produces their startling conclusion that for world supply, considered as a whole, that peak time is right around now. Crucial to their argument is the point that no matter how many more billions of barrels of oil reserves may yet be found in oil exploration worldwide, the length of time it takes to develop such resources means it is practically impossible for this oil to be brought on stream fast enough to offset the decline, now predicted by the bell curve, from our current major producing fields.

It should be noted that practically all oil companies and also many – but not all – oil industry consultants disagree that the peak of daily production will come anytime soon, although they are of course forced to logically concede the concept itself as describing something that must inevitably happen some day. The key grounds for the relative equanimity with which most industry professionals react to the peak oil theory – which none will gainsay in terms of the underlying mathematics of production versus field depletion, or the accuracy of Hubbert's original prediction regarding the US – is that peak oil calculations themselves tend to ignore new forms of "unconventional" supply. (Such as the oil sands mentioned above, various gas-to-liquid production processes, and the burgeoning bio-ethanol industry.) Some of these processes produce actual crude, others produce a different type of hydrocarbon which can nevertheless be used in much the same way as crude – but as far as the oil industry is concerned, these unconventionals are as valid sources for daily barrels of production as conventional crude oil, and should be counted in production growth estimates.

Peak oil advocates are adamant that conventional crude oil production per se will peak in the next couple of years, while an influential book on the subject (Deffeyes' *Hubbert's Peak*) actually specifies 2009 as the peak global production year. At the same time, however, many oil industry economists and consultants tend to see oil supply growing until at least the mid-2020s. Their view is a combination of taking OPEC at its own word – the cartel itself insists that with its current reserves it can continue to grow production for this long – while presuming that non-OPEC supply will continue to grow until perhaps just before 2020. (And, importantly, they also include unconventional supply across both groups: for instance, oil sands production in Canada, and gas-to-liquids production in Qatar). It should, furthermore, be noted that peak oil theory additionally involves making some heroic assumptions regarding the total amount of recoverable, conventional crude oil that will eventually be extracted from the planet through the whole history of the oil industry – statistically-derived assumptions which by definition will never be proven until after the fact.

In energy security terms, the spectre of peak oil mandates a massive shift away from oil-dependent industrial society into alternative sources of energy such as nuclear or renewables. Once the world realises it has moved past peak oil there should theoretically be an unstoppable explosion in oil prices, both to ration demand in relation to falling supplies and to maximise the incentives for remaining production. In geopolitical terms, however, the prospect of peak oil presages unstoppable explosions of a more literal sort, a scramble between countries to secure an ever-dwindling supply of this key commodity – which could all too easily degenerate into new wars.

Resource nationalism

The idea of inter-state competition and struggle over supplies of diminishing key resources, which seems the awful corollary of peak oil, feeds into another theme of the past few years: "resource nationalism". Simply put, resource nationalism is what happens when states realise that it might not be in their best interests to continue to adhere to the precepts of global free markets when considering the disposition of such key natural resources as oil. It is seen as biting on both sides of the supply and demand equation.

On the supply side, governments (like those in Russia and Venezuela) decide that control over the natural resources originating in their countries is such a strategic advantage, in terms of both their own independence and the supply-side sticks and carrots that can be brandished against other less-endowed states, that they move to exclude both foreign and/or private commercial interests from involvement in the sector. Russia's change in subsurface resource laws such that all oilfields over a certain size are deemed "strategic" and therefore must be majority-controlled by Russian interests; its destruction and expropriation of the privately-owned Russian oil company Yukos; and its stripping from UK/Dutch oil major Royal Dutch Shell of its majority control over the flagship Sakhalin II oil and gas project in the Russian Far East, are all good examples of supply-side resource nationalism at work in recent years. To which, of course, we might also add, as another obvious example, the expropriation by the Venezuelan government of the Orinoco belt oilfields, previously controlled by major international oil companies.

On the demand side, resource nationalism is manifested in the state-controlled oil companies of emerging Asian powers such as China and India becoming evermore avaricious in their pursuit of producing assets, whether in terms of companies to be bought, or properties and exploration permits to be snaffled up in licensing rounds. Thanks to the invariably deep pockets that come with state sector ownership, they are able to outbid and squeeze out the Western private sector interests that have traditionally dominated global oil production. State-controlled Chinese oil firms in particular have made prodigious efforts to widen their global footprint. In 2005, CNPC purchased a majority stake in previously Canadian-owned PetroKazakhstan; in 2006, fellow state concern CITIC Group bought up another previously Canadian enterprise operating in Kazakhstan called Nations Energy; and in the same year Sinopec bought 49% of Russian oil company Udmurtneft, while CNOOC bought 45% of a Nigerian offshore block from TotalFinaElf and its state-owned Nigerian partner.

Various arms of the Chinese state-controlled oil complex and Indian state-owned ONGC have even joined forces to secure assets – in 2005 ONGC teamed up with CNPC to buy Petro-Canada's Syrian assets, while in 2006 ONGC and Sinopec joined forces to buy a half-share in Colombian oil company Omimex. Some commentators have therefore been worrying about the emergence of an "axis of mercantilism" between China and India, "mercantilism" being the traditional term used in political economy to describe such state direction of economic activity. But if there was one deal that really highlighted the battle between state-sponsored mercantilism and the free market, it was CNOOC's attempted purchase of privately-owned US oil firm Unocal in 2005. A furore of objections expressed in US policy circles and by US lawmakers caused CNOOC to withdraw its bid before the takeover actually went to an official ruling.

The attitude toward the deal expressed by many in the US was neatly summarised by Richard D'Amato, the US senator heading the United States-China Economic and Security Review Commission: 'The Chinese treat energy reserves as assets in the same way a 19th century mercantilist nation-state would. Their goal is to acquire and keep energy reserves around the world, and secure delivery to China, above and beyond any market considerations. To do this, [they are] willing to pay above-marketplace premium prices in order to gain exclusive control over oil and gas. China believes it can only achieve energy security through direct control of reserves. This hoarding approach directly conflicts with the efforts of the US and other countries in the International Energy Agency to develop fungible, transparent and efficient oil and gas markets.'

Whichever way it is manifested, rising resource nationalism spells a diminution of opportunity in and control over the global oil markets for the Western private sector interests that have historically dominated the business – the integrated oil majors listed across Europe and America like BP, (Royal Dutch) Shell, TotalFinaElf and ExxonMobil. It also, incidentally, offers another twist on the peak oil theory. Some analysts explain that, regardless of what might actually be the correct answer for the year in which existing discovered production is going to peak, and whether or not unconventional supply should be included or not in such sums, in reality the tightening and often hostile grip of state interests on key oil production regions such as Russia and Venezuela means that – even if it might in theory be geologically possible to increase daily production significantly from current levels – in reality this will not actually happen because the necessary investment will simply not be made in a timely fashion. In this view, calling the peak oil moment around now is more a matter of realistic pragmatism than strict geological inevitability.

Twilight in the desert

If you put together all of the real world trends and the suppositions of the related grand narratives recounted above, you might come up with something close to an oil price apocalypse, a perfect storm of geological and political risk factors spelling higher prices indefinitely. What could possibly make things worse? Perhaps the idea that Saudi Arabia, supposedly the last remaining source of discretionary swing supply capable of easing the global oil market, might not even possess the spare capacity it advertised to the world. This was the idea put forward by energy sector investment banker Matthew R. Simmons in his celebrated and controversial book of 2005, *Twilight in the Desert*. His key assertion was that while hard data on the performance of the various giant oilfields accounting for the bulk of Saudi supply was denied to the wider world by Saudi national oil company Saudi Aramco (as a matter of state secrecy), study of an extensive archive of Society of Petroleum Engineers (SPE) reports – spanning decades and documenting the technicalities of the development, commissioning and operation of these same fields – indicated that in reality Saudi Arabia could not reach the production capacity levels of which it continually told the world it was capable.

In fact, according to Simmons not only would it fail to reach its targeted capacity of 12.5 million barrels per day by 2009, but there was a real risk of a catastrophic collapse in Saudi production. As he wrote in 2005:

> With some agile planning and execution of many new projects, Saudi Aramco may well be able to maintain its current production level for some time. It is not impossible that the kingdom may even achieve a marginal increase. But the odds are better that output might go into a gentle and protracted decline. These would seem to be the best-case scenarios, however, and they are vulnerable to a tipping by any number of relatively minor events… A production decline of 30 to 50% in a period of five years or less in any or all of Saudi Arabia's key producing fields is not out of the question.[2]

While the Saudi government refused to engage in debate over these issues, or to allow Simmons or other outside experts to validate or disprove his claims, believers in this Saudi supply "sunset" theory could

---

[2] *Twilight in the Desert*, Matthew R. Simmons, pp.358-9, (Wiley, 2005).

only imagine what such an abrupt stripping-away of the already threadbare OPEC supply buffer would do to oil prices.

1.4 Oil's Unstoppable Explosion

The graph in Figure 3 sketches out the ascent of oil, in terms of daily closing prices on Nymex, from the beginning of 2006 up to its peak in July 2008.

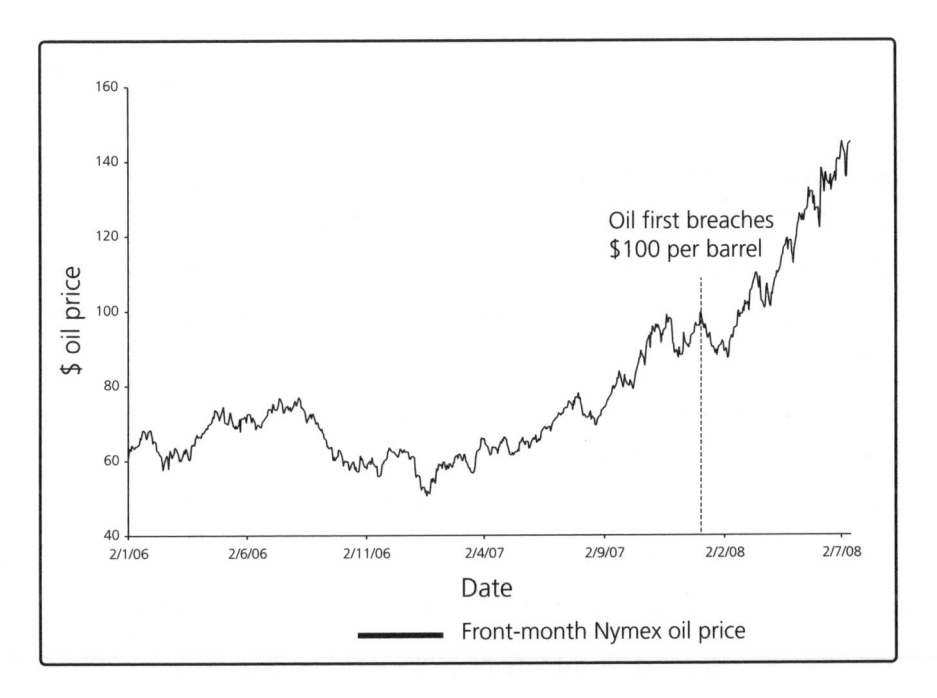

Figure 3: The ascent to oil's peak [Source: Thomson Datastream]

From the first trading day of 2008 through to July 11 (the day oil reached its all-time intraday high of $147.27 per barrel) all of the considerations sketched out above, both apparent fundamentals and grand narratives built on those observations, continued to figure heavily in the discussion of where the oil price was heading and what was driving it there. In attempting to weigh up the perceived importance of various popular memes such as these, it is a somewhat crude but nevertheless well-accepted method to simply count the number of times words or phrases appear conjoined in news reports through a given period. Such an analysis does indeed show that as the oil price grew rapidly from the start of 2008 into that summer, so too did media linkage of this price movement to our collection of fundamental supply and demand factors.

Using the extensive Dow Jones Factiva database of published news articles worldwide, we find that whereas for all of 2007 the use of 'Iran' in the same report as 'oil price' totalled 1,897 articles, for just January 1 to July 12 2008 (we use this date to reflect the fact that comment on oil price movement is often in newspapers a day after the event), 1,557 articles used 'Iran' alongside 'oil price'. For 'Nigeria' and 'oil price', the full-year 2007 count for articles was 1,004, for just January 1 to July 12 2008 the count was 1,335. For 'Venezuela' and 'oil price' the split is 803 articles in all 2007, versus 812 articles in our 2008 period to July 12. 'Peak oil' as a phrase garnered mentions in 179 articles in all of 2007, but 319 in the 2008 period to July 12, and even the comparatively more recherché phrase 'resource nationalism' – a concept which does not lend itself easily to quick-fire news reportage – managed 75 mentions in all of 2007, but 82 in the 2008 period to July 12.

We can also point to some specific sample manifestations of these fears along our oil price appreciation timeline from New Year into summer 2008. In both January and April 2008, US and Iranian naval vessels had some sort of confrontation in international waters in the Persian Gulf. While in both cases stopping short of outright hostilities, each side accused the other of provocation and the incidents raised fears over a conflagration engulfing the key crude oil shipping routes there. In February 2008, the Turkish army entered the autonomous Kurdistan region of northern Iraq in pursuit of Kurdish guerrillas, with the incursion being condemned by the central Iraqi government, and a potential flare-up of violence in previously peaceable northern Iraq seen as a possible threat to remaining Iraqi crude production, half of which is exported via the Kirkuk pipeline running south of Kurdistan into Turkey.

In June 2008 the Niger Delta militants of MEND stepped up their campaign with a gunfire assault by speedboat on the floating production vessel at the Shell Bonga oilfield offshore of Nigeria. While no lives were lost, the incident shut in another 225,000 barrels per day of Nigerian production as Shell was forced to evacuate the vessel. The attack also signalled a new level in disruption. Prior to this, the insurgency had affected onshore and shallow-water facilities around the Delta, but the offshore facilities producing significant quantities from deepwater fields had been seen as immune to militant attentions. Bonga was, however, 65 nautical miles offshore. By taking to the waves, MEND had dramatically punctured this complacency, and dragged whole new swathes of Nigerian production into its danger zone.

Customers around the world were feeling the heat as the oil market reacted to each fresh twist in various supply disruption scenarios, and oil tracked steadily up from $100. By the end of April 2008, US gasoline prices had reached $3 per gallon for the first time ever – then as soon as early June, they had reached $4 per gallon nationwide. In another couple of weeks, drivers in some US states were paying $4.50 per gallon. Meanwhile in the US alone, high fuel costs forced eight small airlines to completely shut down in the period from January to June 2008, with another two such services operating under Chapter 11 bankruptcy protection, while the major "network" carriers such as American Airlines, Delta, Continental, US Airways and Northwest were drastically chopping routes and retiring capacity.

Public outrage at high fuel prices was definitely a factor motivating an ongoing series of hearings by both arms of the elected US Congress, House and Senate, which will be examined in greater detail in due course. However, angry as the mood was in America, the inconvenience of US consumers pressured by high pump prices was not likely to destabilise the government. The same could not be said of developing nations around the world, particularly in Asia, whose governments are in the habit of regulating fuel prices to the benefit of their citizens, particularly their masses of poor people for whom fuel costs are already a disproportionate share of outgoings.

Through caps on fuel prices, these governments effectively subsidise the cost of oil to domestic consumers. As the crude price rose, so did the cost of this subsidy – to breaking point in some cases. In late May 2008 the governments of both Indonesia and Malaysia, neighbouring countries, announced roughly 30% hikes in fuel prices. Newspapers in both countries quoted expert opinion fearing civil strife as a result, and in Malaysia opposition parties called for rallies against the step. Around the same time, India raised its regulated fuel prices by around 10%, Taiwan by some 13% and Sri Lanka by 24%. *The Economist* magazine commented at the time: 'Across the emerging world, governments fear that lifting fuel prices will hurt the poor and so trigger social unrest.' But given the ever increasing drain on their coffers which maintaining these subsidies meant, many felt they had no choice.

Meanwhile a chorus of financial analysts kept hammering out the message that these high oil prices were completely justified as reflecting the constraints on supply versus galloping demand. Investment banks, advising clients what to do with their money, were at the forefront – and perhaps none more so than Wall Street titan Goldman Sachs. At Goldman a team including their global head of commodity research Jeffrey Currie, in London, and US-based oil equity analyst Arjun Murti, had already established a reputation for prescience in making very bullish calls on the oil price. They had initially surprised many with their predictions, but tended to be correct: it was in 2005, when oil was actually trading around $50 per barrel, that Murti first predicted a 'super-spike' in the price to $100.

From autumn 2007 Goldman had been saying oil was likely to streak past $100, and now the bank was again forecasting higher prices. In early May 2007, Murti commented in a research note that 'the possibility of $150-$200 per barrel seems increasingly likely over the next 6-24 months'. Later the same month, Currie's research team revised upward their estimate for the price they expected oil to average through the second half of 2008, moving from $107 to $141 per barrel. It would however be very unfair to single out Goldman alone on this issue – other investment bank analysts at houses such as Barclays Capital and Merrill Lynch were equally confident that prevailing prices were justified by fundamentals, and agreed with Goldman that it would not be until persistently high prices caused genuine "demand destruction", in other words forcing the global appetite for oil back into line with the actual supply-side realities, that prices would moderate.

By this point, however, even confirmed believers in $100-plus oil were becoming discomfited with some of the wilder predictions being made. Following wide publicity attending Murti's $200-per-barrel statement, Paul Horsnell and Kevin Norrish of Barclays Capital commented pithily in one of their own research notes that there seemed little point in adding further to ever-higher round numbers, with oil price forecasts being thrown around like so much 'analyst bling' (to use their memorable phrase).

1.5 The Dissidents

Not all commentators were convinced of the case for triple-digit oil. Among investment bank analysts one or two, perhaps most notably Ed Morse and his team at Lehman Brothers and Colin Smith of Dresdner Kleinwort, had repeated through these months up to early July 2008 their opinion that oil prices above $100 were essentially a 'bubble' blown by speculators, in the same sense as described in our introduction. We shall examine the whole concept of a "speculative bubble" at length, but suffice to reiterate for now that labelling asset or commodity price inflation a *bubble* is to recognise it as unsustainable and liable to *burst*, or collapse. This is because the price inflation is caused more by speculative pressure from investors hoping to profit by simply riding upward price momentum than it is due to genuinely fundamental commercial factors.

Morse and Smith stood out among their peers, but they were certainly not alone in fingering speculators as a major force driving the oil price up. Lined up alongside them were a good number of oil industry professionals themselves, including company executives and energy sector consultants. Most publicly, OPEC itself was also loudly insisting that speculators rather than fundamentals were to blame for oil nearing $150. At the International Energy Forum in April 2008 (a meeting in Rome between major oil consumers under the aegis of the IEA and major oil producers, with OPEC well-represented), Saudi oil minister Ali Al-Naimi had bluntly stated that, 'Speculation in the futures market is driving prices. Today, there is no link between oil fundamentals and prices.'

Nevertheless, the cartel was under a lot of pressure, not least from the US government, to do something to alleviate the situation. When then-US president George W. Bush visited key producer Saudi Arabia in early May 2008, the Saudis themselves undertook to add a further 300,000 barrels per day to production. A month later they pledged another unilateral increase of 200,000bpd – a total of half a million barrels per day added into fundamental supply and demand calculations in short order. Incredibly, this had seemingly no impact on oil's inexorable price rise – yet at the same time, the Saudis noted they were already finding it hard to sell all their output! Later in June – with the prospect of the promised Saudi increases having not reduced oil prices at all – OPEC president Chakib Khelil made a similar point when justifying no fresh production increases by the cartel beyond the latest Saudi commitment. Khelil said that other OPEC countries such as his own state Algeria were also finding it impossible to sell all their production. Representatives from Qatar, Iraq and Libya likewise affirmed around this time that they were producing more than consumers required from them at that moment.

The Saudi government and Khelil did indeed both acknowledge strong demand for crude types preferred for diesel production – which in various ways disadvantaged their own supply, with the Saudi oil being too sour and the Algerian being so sweet it was preferred instead for gasoline production (since it gave such a good yield of this fuel in particular). Nevertheless, the OPEC front argued publicly that with the market as a whole so obviously well supplied, it could only be speculators responsible for oil price inflation. Evaluating arguments over the effect of speculators and the objections raised in turn against them, however, requires an understanding of how the global "oil price" is actually determined. Welcome to the world of "paper barrels".

2

Paper Barrels

'What happens in the futures markets does not stay in the futures markets...'

Testimony of Michael W. Masters, managing partner of Masters Capital Management, United States House of Representatives June 23 2008

2.1 Whose Oil Price?

Asgard, Bonny Light, Forties Blend, Louisiana Light Sweet, Premium Albian, Terra Nova, White Rose: as instantly evocative as cigarette branding, these are in fact the names of some of the 160-odd commonly recognised "benchmark" crude oils produced and traded around the world. In a roster running from Abu Bukhoosh to Zueitina, each benchmark is defined by a list of attributes that crude oil should meet if it is to be sold under that name – primarily chemical qualities such as specific gravity, sulphur content, pour point and mineral traces. As crude oil itself is a naturally occurring resource with infinite small chemical differences in product from field to field, invariably the benchmark name is also a guide to a narrowly defined point of origin for the oil in question. Abu Bukhoosh is locally-produced oil sold out of the port of the same name in Abu Dhabi, Zueitina is likewise named for its port-of-sale in Libya, while Asgard and Forties refer to complexes of oilfields in the North Sea from which these blends are produced by, respectively, Norwegian oil company Statoil, and British company BP.

Benchmark crudes exist because oil refiners need to know what sort of crude oil they are buying to refine into usable products in terms of the heavy/light and sour/sweet distinctions. Ultimately, the attributes of a particular crude may limit the products that can be derived from it through the refining process, for which the full spectrum of output runs from liquefied petroleum gas (LPG), through petrol/gasoline, kerosene/jet fuel, diesel, heating and other fuel oils – to lubricating oils, waxes and even asphalt, plus a slew of precursors and feedstocks for petrochemical and subsequently plastics production, such as ethylene. The crude benchmark system indicates differences between crude oils traded between producers and refiners that in practice merit differential pricing with regard to their refining potential. When production from a particular new oilfield is brought onstream, it may be determined as already comparable enough with an existing benchmark, or capable of being made so through blending – and indeed, many benchmarks themselves are already blends of crudes in any case. Alternatively, the new production may perhaps be reckoned worthy of its own new benchmark and marketed as such by its producer, as Statoil did with Asgard from 2006 onwards.

These pricing differentials obviously require some price to be picked as the basis for others, and the prices of three blends of crude in particular emerged in the 1980s as the prime references or "markers" for the global physical oil trade. These are West Texas Intermediate or WTI, Brent and Dubai. WTI is a particularly light, sweet crude oil that is very easy to refine and is produced in the very region of the United States that the name itself indicates. Brent, produced from the UK North Sea, is another light sweet crude blend and, although slightly inferior to WTI, is the most widely-referenced price for the physical oil trade worldwide. Dubai is produced in the United Arab Emirates and is also light but significantly sourer than the other two benchmarks. Because it is the blend most readily available in the Persian Gulf region – not quite the same distinction as being the most prolifically produced! – Dubai has become the reference for crude sales into the Far East market, as opposed to the Atlantic Basin market dominated by WTI and Brent.

WTI and Brent have wider recognition than Dubai, and traditionally the quality differential sees Brent priced a couple of dollars cheaper per barrel than WTI.

The wild appreciation in the per-barrel cost of oil from early 2007 into 2008 has already been described in terms of the crude oil price quoted on the Nymex exchange in New York. This is appropriate, as it is indeed the price per barrel quoted on Nymex that is the most commonly followed benchmark price worldwide. The New York exchange is seen as the largest and most liquid oil market in the world, hence the best suited to price discovery in terms of matching large numbers of buyers and sellers to reach an optimal price, a price based on the widest possible spread of informed opinion. Yet as described above, in the actual day-to-day trade of physical oil there are many prices that could be quoted for a barrel depending on where it comes from, and even in terms of the predominant reference prices for determining a wide spread of subordinate prices we have perhaps three, and certainly two, benchmarks to choose from. So which of these physically traded benchmark crudes is the actual oil changing hands on Nymex as the "oil price" on which so much attention around the globe is daily focused?

Strictly speaking, none of the above. On the one hand, it is certainly the price for a WTI-standard barrel of "light, sweet crude" that is taken as the global benchmark; however, the price the financial press invariably quotes in that respect is not for a physical barrel of oil available for delivery any time in the next few days, a so-called "spot" or "prompt" delivery. This is verified by looking at any number of online market data feeds, which will quote separate prices for both WTI "spot" at Cushing, the major US oil terminal and trading hub situated in Oklahoma, and the Nymex crude oil contract. It is the Nymex price, usually some cents ahead of the Cushing spot price, which is repeated in daily news reports around the world. But Nymex is not a physical oil market at all, rather a "futures" market, where financial contracts rather than tangible goods are exchanged. The oil barrel priced on this market is at root a virtual construct – a "paper barrel".

2.2 Future Perfect

Futures markets took off in the United States from the middle of the 19th century, and originally grew out of trade in agricultural rather than energy commodities – Nymex itself has been operating for 135 years. Simply put, a futures market is an exchange on which are traded contracts guaranteeing a set price to be realised on delivery of a particular commodity at a specified future date, which could be as close in time as a few weeks away or as far out in time as many years hence. Market participants are said to "buy" or "go long" on these contracts if they wish to take delivery at the quoted price, or "sell" or "go short" on these contracts if they wish to make delivery at the quoted price. As with any market, quoted prices for the good on offer – in this case, forward-dated commodity delivery obligations – change from moment to moment with market perception of the risks and rewards attendant on holding that good.

The "light, sweet crude" oil futures market on Nymex is organised into contracts, each contract or "lot" implying liability for a thousand barrels of crude oil, with the delivery stipulated as being of WTI-standard oil at the Cushing oil terminal. The contracts are dated in terms of specific calendar months, e.g. December 2012 or July 2014, with quoted maturities currently stretching out through some eight years of potential future delivery dates. The contract maturities available are determined by trader appetite, and have steadily lengthened over time – in the early 1990s, the vast majority of oil futures traded had a maturity measured in months alone, and growth in contracts longer than three years did not begin until 2004. The set of prices for ever more distant delivery dates which is generated in the day-to-day trade of such contracts is called the futures "strip". The same strip represented as a price graph extending along an axis of increasingly distant calendar dates describes what is known as the futures "curve", although it may sometimes be another shape entirely. The Nymex contract everyone follows for the global benchmark oil price is the so-called "front-month" contract, which is the calendar contract right at the front end of the curve and hence closest to expiry of trading at any given moment. The curve runs from this front-month out to the furthest point of the "long-dated" maturities. Figure 4 sketches two sample oil futures curves from two separate dates in recent years.

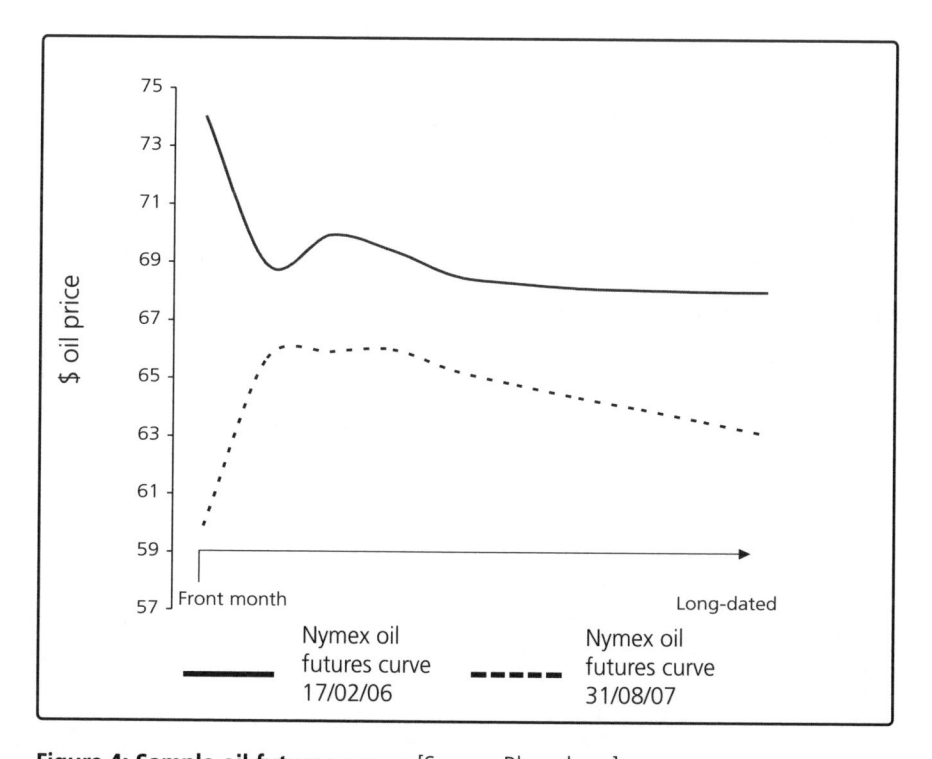

Figure 4: Sample oil futures curves [Source: Bloomberg]

As all futures contracts have a fixed delivery date, all eventually pass through an expiry date beyond which they cannot be traded on the exchange, at which point outstanding contracts held by market participants move into the physical delivery phase. Trading expiry is generally set a couple of weeks prior to stated delivery – this period being seen as reasonable for real-world logistical arrangements to be put in place for the delivery, which is ultimately presumed as a possibility for *all* contracts. Officially, Nymex oil contracts cease trading on the third business day prior to the 25th day of the month prior to the actual delivery month. This means that already in early June, the front-month Nymex price will be referred to as 'oil for July delivery' – and once the date passes beyond the third business day prior to June 25, this will shift to 'oil for August delivery', even though we might not yet be in July! Likewise at this point, July becomes the "prompt" or "delivery" month rather than the "front-month", even though we might not yet actually be in it.

As noted, the front-month Nymex light sweet crude contract is generally very close to the prevailing price for actual spot delivery of the same standard of WTI crude at Cushing, and this is precisely because of the trading and delivery timetable for front-month crude. By the time trading expires for any given front-month contract, there is something in the order of a week or two before delivery at Cushing falls due. Meanwhile at the point the previous front-month becomes the current delivery, or "prompt" month, there is never more than around four weeks before the new front-month in turn becomes the next delivery month. So logically, the difference between the cost of front-month Nymex crude and spot delivery at Cushing is the fee you would pay for keeping oil you could buy spot at Cushing, right now, in its storage at the same terminal for the roughly two-to-four-week period until that next front-month becomes the "prompt" or actual delivery month –

and that is not very much per barrel after all, generally a matter of cents. And yes, as logic would dictate, the difference between the two prices is normally greater at the start of a new front-month and less as it nears expiry.

The utility of such a futures market lies in it allowing both the producers and the consumers of a particular commodity to fix in advance the price they will either receive or pay for it, to "hedge" this price risk as the technical term goes, thus allowing them a certain level of certainty in organising their business. For example, a small oil company might be trying to decide whether or not to take out a loan to fast-track development of a marginal, low-yielding discovery somewhere in the back-of-beyond in Texas. To make it worthwhile the company needs to know that the price it is going to get for the oil produced from this well in the future – in, say, nine months' time – will generate sufficient payback on the investment, including all sunk capital plus the operating costs (including the interest on the loan) plus whatever rate of return the developer judges it needs to get by.

The oil company can try to hedge this price risk by going to the futures market and checking if the prices currently quoted for deliveries of oil, across the calendar window through which the well will be producing, are high enough to justify this course of action. If the quotes offered on the market are rich enough, the company will sell contracts at that price through a futures broker – thereby locking in profitable purchase for its production when it is pumped from the well nine months hence. The oil company is of course free, through its futures broker, to quote its own price for selling contracts at whatever level it wishes, rather than simply taking a price from an established market maker. However, if the price it requires is always higher than that offered by others seeking to sell future deliveries, it will never get a deal done.

There is another side to the equation. Looking at exactly the same nine-month forward quote for oil in the futures market might be a small oil refinery on the Gulf Coast of the US. Say they were seeking to ensure adequate supply of crude input through what management fears might be a disruptive storm season, in the same window the oil company is selling into. The refiner would prefer to guarantee delivery of sufficient crude at an agreeable price through this window rather than tie up capital in buying the equivalent stock now and then also incurring storage costs on it through the intervening months. The refiner can go to the futures market, and it if likes the price on offer for crude in nine months' time, it will buy sufficient contracts to guarantee the deliveries it wants.

Fast forward nine months, and it is possible that the storm season has indeed been a real humdinger – with multiple offshore platforms out of action in the Gulf of Mexico, the price of oil for spot delivery has rocketed. Tough luck for the oil company, which sees its rivals pocketing much higher prices for their crude while it receives the lower price it locked in by selling the futures contracts. But great for the refinery, which has guaranteed deliveries of crude at the same lower price the oil company has sold at, while rival refineries are paying through the nose for whatever they can get their hands on. Nevertheless, the oil company will console itself with the thought that the storm season could have easily been a damp squib that saw oil prices dive below the price it sold at – and it still receives the funds it calculated as sufficient for its own profitability.

This is a key point to grasp about the utility of hedging for companies that are genuine physical producers or consumers of the commodity in question. At the end of the day, any gains or losses incurred on their futures transaction (with regard to what they could have bought or sold at on the actual date) should be seen as secondary to the fact that they are still buying or selling as they originally planned to, at the price locked in through the hedging operation. The risk of hedging at what turns out to be inappropriate levels is the flipside of locking in now the price the commodity will be traded at in the future, and it is a risk many physical producers and users of a commodity are happy to take for the obvious benefits in terms of cash flow forecasting. Furthermore, bankers and other financiers will often insist that a producer or user of a particular commodity hedges this price risk at levels sufficient to guarantee the repayment of interest and principal on any debt they may provide to such a business.

In a sense, genuinely physical hedgers theoretically *cannot* "lose" on a futures trade, as by definition they will be getting precisely what they wanted at a price they were happy enough to agree to previously. Complaining that the other guy paid/received less/more for the same commodity on the actual day of delivery because they had not hedged, or had been able to hedge at a different price, is, in the purest version of hedge theory, beside the point.

2.3 Future Imperfect

So much for what we might call the Arcadian, pure theory of hedging, in any case. The reality is more complicated – because 99% of futures contracts on Nymex do not actually run to physical settlement of the trade through actual delivery, even when the parties involved are physical producers or consumers of the commodity. They are instead settled through the practice of "offsetting" – or cash settlement of the difference between the price for the calendar delivery previously agreed by the market participant when buying or selling the original contract, and the price for the same calendar delivery at the particular moment the market participant is seeking to settle.

By definition, this will not be the same as the actual price their commodity may trade for on the actual specified delivery date, as the offsetting has to take place prior to expiry of trading in the contract in question. And this is because offsetting is achieved simply by transacting on the futures exchange the equal and opposite trade to the original trade being settled. If a trader went long of 50 December 2012 contracts, he will "offset" or settle his position by going short of 50 December 2012 contracts, or vice versa.

Offsetting works because of one of the key features of a futures exchange that makes it such an attractive proposition for managing price risk in the first place. Exchanges such as Nymex radically reduce the risk of doing business in contracts for future delivery by taking on the settlement obligation themselves, with the entity of the exchange itself becoming the counterparty for every transaction concluded by contracts under its auspices. That is, if someone holds a contract from the exchange, then the exchange itself both guarantees and takes charge of fulfilling it on the due date, regardless of whatever happens to whichever party was actually quoting the price at the time the contract was taken up.

Futures markets such as Nymex cover the risk this guarantee could pose to their own financial stability primarily by collecting a good faith deposit from each party trading on the exchange, to cover at least a portion of the financial exposure inherent in their contract commitment (such a deposit is called the margin); by collecting additional fees from certain classes of exchange member; and by insurance contracts. And indeed, because as noted above actual full payment under any contract held by a market participant is only due upon physical delivery, it is instead only this margin that any participant actually pays into the exchange on the actual day a futures position is established, if they are buying. Occasionally the margin required will be equal to the whole value of the futures transaction; in which case the contract is said to be "fully collateralised".

This blanket settlement guarantee, which the exchange extends over all its contracts, means it is actually indifferent as to whether the market participants originally taking each side of a contract struck at a particular price and time actually both remain on the hook for that particular obligation, right up to physical delivery. After all, the exchange itself does not trade contracts on its own account. In both the selling and the buying of the contracts, any offsetting party has had to strike deals on the basis of quotes offered by other real market participants. That is, if someone is buying a delivery obligation, they can only actually do it if someone else is willing to sell it; and if by extension that same buyer then wants to sell *out* of their existing obligation to take delivery, they can only do it if someone else is bidding to buy *into* that same dated delivery obligation.

In the latter case, the net position of counterparty exposure the exchange holds overall does not change, merely the identity of the participants involved in constituting this exposure. Meanwhile the net delivery obligations between the exchange itself and the original buyer have accordingly been reduced to zero. First the buyer took a position whereby they were obliged to take delivery of a set amount of the commodity at a certain contract date from the exchange, but subsequently they took out a position whereby the exchange was also obliged to take delivery from them of the same amount of the commodity on the same contract date. So now there is no need for any physical commodity transfer whatsoever between this market participant and the exchange. There is, however, a financial settlement to be concluded, depending on the difference in contract price between the two dates the original buyer put these two opposing positions in place.

Returning to the example of the refinery. If it originally bought the oil nine months out at $50 per barrel, then by the time actual physical delivery date is looming, the oil price for delivery on that same date might have jumped to $80. If the refiner moves to close out this position through offsetting rather than taking actual physical delivery, it will have bought a quantity of contract obligations at $50 per barrel and then later sold the same contract obligations on for $30 more per barrel. So as the universal settlement counterparty on all contracts, the exchange now owes the refinery that $30 per each barrel. Upon settlement by offsetting, it will duly credit the refiner's account with those profits. Where has that money come from? Ultimately from whichever market participants have been holding the other end of those contract obligations elsewhere on the exchange.

While it may indeed be the exchange itself that has paid out the refinery its profits on offsetting, it is not a charity. It meanwhile has rights over the margin accounts of the market participants holding the other side of these positions – which may not actually be the original seller of the contracts, if they too have already decided to close out their position by offsetting. Returning to our small oil company, Texans are canny people and it may not have been too long after selling their oil nine months forward at $50 per barrel that these oilmen realised events in the real world were moving against them. Perhaps weather scientists are already predicting a bumper storm count in the Gulf of Mexico. By the time there are only five months left until contract maturity, oil for delivery on that date is already trading at $65. This is bad news for the Texans. If this trend continues, by the time delivery falls due they could end up selling their product for far less than they might otherwise be able to. And while in our "pure" hedging theory above, hedging companies are supposedly prioritising the security of a set amount of guaranteed cash on a certain date – over the possibility of getting more cash on the actual date in question – in real life, no one likes to leave money on the table.

The Texans decide to exit the position by offsetting. They originally sold contracts for future delivery at $50 per barrel; now they are buying back the same amount of contracts for future delivery at $65. Their delivery obligations cancel out, but by selling low and buying back higher they have lost money in achieving this outcome – $15 per barrel. This is the amount the exchange docks from their margin account upon settlement by offsetting. It is only half the amount the refinery eventually receives in the example above, but bear in mind the refinery has not itself offset out of its position yet. If it were to offset at the same moment as the Texans, five months before rather than just prior to delivery, then the price would be $65 and the refinery itself would only have made $15 per barrel – exactly the margin amount the Texans have now forfeited with the exchange, which is precisely why the exchange would be able to pay out the refiner in turn.

In our example the refiner is, however, actually still holding on to their position at five months out, even as the Texans sell out. By definition, as the Texans are buying themselves out of an existing obligation, whoever is in turn selling them this contract on the exchange has now taken on the other end of the refiner's original transaction. This new party – let us presume an actual physical oil trader which has just bagged the rights to a consignment of oil, set to be loaded on a tanker for the US in a couple of months – has sold this oil for delivery in five months at $65. By the time the refiner comes to offset its own position when oil for the same now-imminent delivery date is $80, just prior to delivery, this oil trader will now also notionally be down $15 versus the exchange itself.

Yet the oil trader itself does not need to offset their own position for the refiner to get paid the full $30 it is owed. The exchange can pay the refinery the $15 already realised from the Texans, and the other $15 it can pay from its own massive pool of funds precisely because it has security over the oil trader's margin account, even if this party has not yet crystallised its own gain or loss versus the exchange. Throughout the infinite permutation of offset possibilities – as market participants trade in and out of positions – the exchange will always balance money paid out to winners, on offsetting, with margins liquidated from losers when they close out their position. Particularly as, in a real futures exchange, traders nursing a notional loss at prevailing contract prices face "margin calls" to keep their margin accounts topped up sufficiently to cover this loss, regardless of when or whether they actually intend to crystallise it through offsetting.

Why do the overwhelming majority of market participants prefer to settle their futures exposure through offsetting without letting these positions run to physical delivery? Simply because, for many, making or taking physical delivery of the commodity in question on the due date is impractical. The refiner might actually be able to buy oil far closer to its operations than Cushing; the oil company may have a standing marketing agreement with a preferred trading party through which it sells all its oil. A physical commodity user may also not be able to let their future contract run to delivery because the actual oil they are seeking to sell or to buy is not actually the exact same standard as that stipulated in the contract. An oil producer may be selling a slightly different blend of oil from WTI, yet one which is still priced in relation to WTI.

Yet in either case – the market participant not wanting to meet delivery conditions for its own convenience, or not actually being able to by virtue of their seeking to trade a different grade of oil in the physical world – the hedging operation settled by offsetting has nevertheless had its intended economic effect. Having exited its obligation to take physical delivery of oil at Cushing for $50, the refiner will instead buy it on the open market at a trading hub far closer to its base of operations. As we know, the spot price there should actually be near enough to the $80 per barrel the refiner sold out of its contracts at, just prior to expiry of trading – but will inevitably not be exactly the same. Perhaps the refiner ends up actually paying $82 per barrel – still, having already realised $30 per barrel on the futures exchange, its effective price for that same quantity of barrels has been reduced to $52, near enough the $50 originally aimed for in any case.

Likewise perhaps the oil our unlucky oil trader has in his tanker is not WTI, but normally sells for a $4 discount to the benchmark. If by the time he lands his cargo on the Gulf Coast the spot price (the price his real-world trading counterparties in the physical market will be referencing in calculating his payment) is $82 per barrel, he will receive $78 per barrel. As he is already down $15 to the futures exchange after offsetting out of his trade at $80 just prior to docking, his actual realised price is $63. Again, near enough but not quite the $65 specified with the hedge – but then whereas our trader always knew his actual sale price was going to be discounted by some $4 to benchmark WTI, the small increase in spot prices between him closing the futures position and selling his oil has closed this gap by $2. This time around the combination of pricing and temporal mismatches between closing the futures trade and realising the physical trade has worked in his favour.

In the real world, there are many niggling little factors – including the exact hub physical oil is being traded through, the differential it may price at compared to the hedge-able benchmark, and the risk of this differential itself changing through time – which mean that the exact outcome of a hedging operation may not be exactly what was intended.

Imperfection will out, at the end of the day; but the difference will normally be small and is referred to as the "basis" risk, seen as the residual and essentially irreducible risk of doing business in such markets, even after putting suitable hedges in place.

Even allowing for basis risk, through offsetting in the futures exchange oil market, participants can achieve their originally intended economic effect without even having to physically deliver the commodity to the exchange-specified location on the specified date. That is the real beauty of futures markets – and, some would also say, the curse. Because the flipside of all of the above is that if it is possible to recognise a gain on a futures exchange through the practice of offsetting, while never actually having to either make or take physical delivery of the commodity in question, then by extension it is possible for these gains to be realised by futures market participants who actually have no real world business interest in the physical commodity in question at all. This is where the speculators come in.

2.4 Enter the Speculators

The word "speculator" itself can mean different things to different people. However, as far as futures markets are concerned, a speculator can be handily defined as a party 'who does not produce or use a commodity, but risks his or her own capital trading futures in that commodity in hopes of making a profit on price changes' (this definition is in fact taken from the Nymex website). Speculators make money by correctly guessing the direction the price of a commodity will actually develop in over time, going long at the price currently available for a contract if they expect it will rise, or going short at the price currently available if they expect it will decline.

Offsetting the original exposure after the price has moved in their favour then realises the difference manifest over the period. For physical hedgers, this difference feeds into another, real world transaction, as a balancing mechanism to ensure the price originally locked-in is actually realised. The refinery needs to take the $30 per barrel from the futures exchange it gained – from originally buying a contract at $50, and then liquidating once the same contract price hit $80 – to offset against the (roughly) $80 per barrel it actually ends up paying for its oil to achieve the (roughly) $50 per barrel it originally aimed for. For speculators with no interest in the actual physical oil market, however, the $30 gain recognised on the exchange if they carried out the same transactions would simply be pure profit.

In a key sense, futures markets actually need speculators. A market consisting only of actual producers or consumers of a commodity might not be liquid enough to enable all parties to hedge themselves at acceptable prices, but with speculators coming into the mix there are more parties to trade this price risk with, and more chances of finding a counterparty for a transaction. One class of speculator in particular is core to the functioning of traditional futures markets: the "floor traders" or "floor brokers", who commonly take the lead in actually bidding and offering the price in open outcry.

Among such market participants are found the "market makers" in a particular contract, professional traders who spend every day on the floor of the exchange as their job and are willing to buy and sell lots of contracts with all other participants. Ideally, they hold any one exposure on their books for as short a time as possible, and seek to make money primarily out of the spread between bid and offer prices they quote, because after all no one can expect a genuine market-maker in a contract, capable of both buying and selling alike, to perform both functions at the same price, or they would never make any money.

The supposed benefits of speculative interest in a futures contract do not, however, stop with the role of market makers. In conventional economic theory, the more information embodied in the market, the more optimal the price discovery for all concerned. Speculators might not have need of the actual commodity, but they may well have some other form of specialist knowledge. They may, for example, be expert in statistical probabilities regarding weather patterns. And this therefore feeds into the pricing equilibrium, and should mean that such risk is already "discounted" in the current price.

Inasmuch as a speculator with such knowledge expects a bad storm season, they will be willing to quote or seek a higher price than others might do when seeking to buy a future delivery of oil. Conversely, if they expect a damp squib they will be happier quoting or taking a lower price than others when seeking to sell a future delivery. Their quotes will take their place within the stack of quoted bid and offer prices from which the market price emerges, helping to form this price regardless of whether or not they actually ever hold physical possession of the commodity. More participation in a market ideally means more information implicit in its achieved pricing – and hence supposedly a more "efficient" price discovery for the commodity in question. Speculative participation is theoretically as welcome in this regard as participation by physical commodity players.

This weight-of-information argument is precisely why the Nymex contract for front-month light sweet crude delivery is looked to as the global benchmark for the oil price. It is the most widely-traded oil futures contract in the world, hence supposedly the most effective for price discovery. Generally, the volume of front-month contracts for WTI outstanding on Nymex at any one time is more than twice the volume of front-month contracts for Brent crude outstanding on the ICE Market in London, the latter being the second most popular oil futures contract traded. So it is the Nymex front-month price which the world prefers to take as its starting point, even though in the real world more physical oil is actually priced with reference to Brent than WTI.

As noted, the front-month Nymex light sweet crude contract is generally very close to the prevailing price for actual spot delivery of the same standard of crude at Cushing. Importantly, however, the price transmission is seen to work from the futures price to the spot price rather than vice versa. That is, the spot price is determined by looking at the front-month contract price and then discounting appropriately. This is what energy sector hedge fund manager Michael Masters meant when, during his testimony before a June 23 2008 hearing in the US Congress on the oil price, he made the comment quoted at the start of this chapter.

A fuller rendition of that statement by Masters is:

> In the present system, price changes for key agricultural and energy commodities originate in the futures markets and then are transmitted directly to the spot markets. For these commodities, what happens in the futures markets does not stay in the futures markets, but is felt almost immediately in the spot markets. Physical commodity producers and consumers trust and rely upon the price discovery function of the commodities futures markets to accurately reflect the overall level of supply and demand, pricing their spot market transactions directly off the applicable futures price.

The spot price of oil, the actual price paid in the real world by physical users of the commodity, is determined by the price of a notional barrel traded in a futures contract obligation, which 99% of the time will be liquidated prior to delivery. It is not necessarily attached to any particular, physical barrel of oil – it is a "paper barrel".

2.5 Growth in Paper Barrels

Prior to the 1980s, there was not much need for active oil price hedging in the US in particular, as the cost of domestically-produced oil was actually regulated by the government. An increasing share of imports from abroad in US oil consumption, and the volatility introduced into international oil prices through events such as the OPEC-inspired oil shock and the Iranian revolution, made this system of regulated prices impossible to maintain. From 1981 the US oil price was free-floating. It is from this date that trade in oil futures on Nymex started to blossom, first sustained primarily by North American interest but, as volumes grew and the pricing achieved credibility, then by market participants from all over the world.

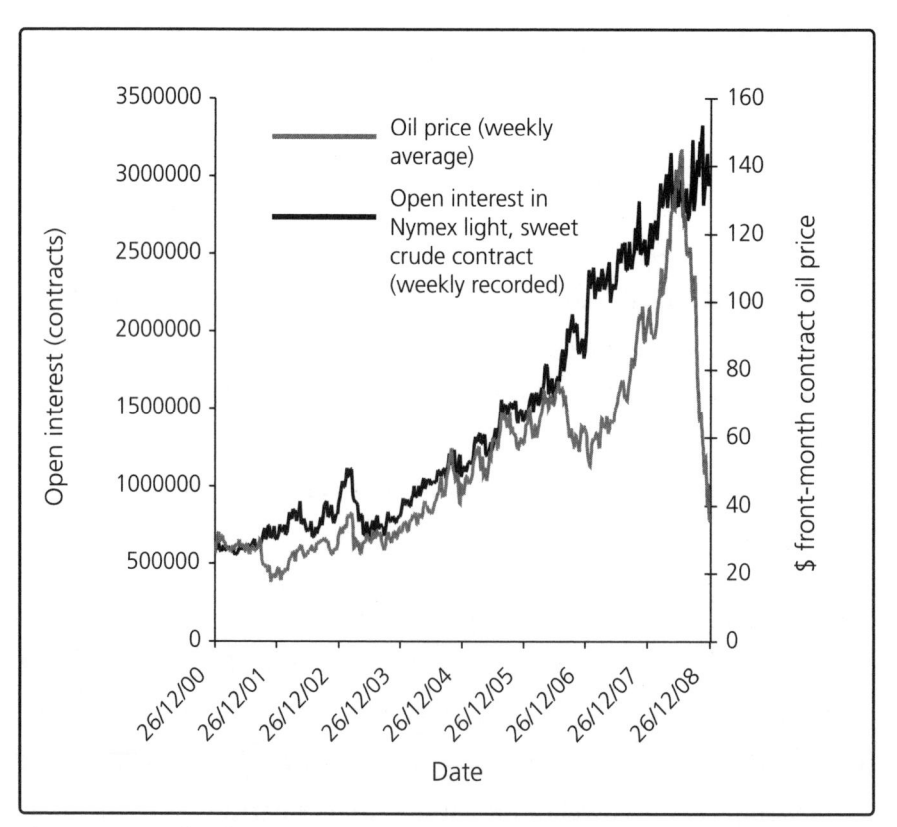

Figure 5: Growth of Nymex open interest [Sources: CFTC, Thomson Datastream]

Figure 5 shows the growth in oil futures trade on the Nymex market just since 2000, expressed in terms of recorded weekly "open interest" across the whole futures curve for light, sweet crude oil contracts traded on the exchange. For comparison, it also shows the path of the oil price itself (as determined by the front-month Nymex light, sweet crude contract) through this same period of growth in open interest. Open interest is the most commonly used metric to refer to the size of a futures market, and references the number of contracts extant for delivery at any one time, i.e. not including offset contracts within the reckoning. As such, it is not quite the same as a simple trading volume measurement – and here is our introduction to the sometimes counter-intuitive nature of the *additive rules* of open interest.

In the beginning was the contract; and the contract had two sides; and the exchange saw that it was good. We can imagine some pristine first dawn of futures trading on a particular exchange, the moment when its first ever deal, for just one contract, is struck by two market participants for a particular date. At that point, "open interest" on the exchange is one contract, a contract by definition having two ends: one end being the single buy obligation incurred in trade to date, and the other end being the single sell obligation incurred in trade to date. Not long afterwards, another deal is concluded between two other market participants for exactly the same date, and open interest in that particular contract maturity has now climbed to two contracts.

Imagine, however, that when the next punter comes to the exchange to buy a contract for delivery on that exact same date again, there is actually no one else likewise newly-come to the exchange seeking to sell that obligation. The punter is in luck, however. One of the buyers in the previous deals is now already unhappy with the obligation they have incurred, fearing they will no longer have use for the commodity in question on the delivery date. They are willing to sell *out* of this delivery obligation to the newcomer so that the latter will take it over for them, and the latter is willing to do so. A third deal is agreed. So open interest on the exchange is now three contracts? No, it remains at two contracts – because, despite the participation in three deals to date on the market by five parties, they have between them only dealt across two whole contracts. Open interest would only have grown to three if two newcomers had struck a brand new deal, a new contract sale and purchase, between themselves, rather than a newcomer facilitating someone selling out of a previously dealt contract.

Contrariwise, open interest on our newborn exchange could actually have shrunk back to just one contract on this third transaction if, instead of having to wait for a newcomer to turn up willing to buy him out of his obligation, the antsy buyer was seeking to offset just as the seller on the other existing outstanding contract also decided it was in their interests to abandon their own position. The antsy buyer is seeking to sell the same contract that this latter participant is seeking to buy at the same moment, both of them seeking to reverse their earlier trades. They are able to agree a price on the exchange and a deal is done. As in this case it is both an existing buyer and an existing seller offsetting their way out of their obligation, one whole contract is liquidated and open interest drops from two contracts to one. The remaining single contract obligation on the exchange to deliver the set commodity at the set date is now theoretically composed from both "other halves" left behind by these two exiting parties – one of course still being a seller of the commodity, the other still being a buyer.

So, with the proviso that open interest does not necessarily reflect the actual absolute volume of futures trades through a period (but rather the net outcome from this trading in terms of still-extant contracts for delivery at any one point), our graph nevertheless indicates a six-fold increase in interest across Nymex oil futures since the start of this decade. This open interest is technically "futures-equivalent" open interest, because it also includes the positions implied not just by straight futures but also by another instrument widely traded on Nymex alongside these contracts: options. Options come in two forms, "puts" or "calls". A put option gives the buyer the option, but not the obligation, to sell the commodity in question at the price and on the date specified in the option itself – which might be, for example, December 2012 oil at $100 per barrel. A call option instead gives the buyer the option, but not the obligation, to buy the commodity in question at the price and on the date specified in the option itself.

Options are more complicated than futures. For a start, they do not necessarily require offsetting or physical delivery, but may instead expire worthless. If, for example, a market participant had purchased a $100 per barrel December 2012 put option, but by this date oil is actually trading at $120, they would be stupid to actually insist on exercising their option. It is said to be "out of the money", and will not be used at all. This is indeed generally the intent of a person selling or "writing" an option: to realise a fee on selling the option itself, but then hopefully never have to make good on the implied obligation (the fee generally being determined by incredibly complicated maths, and supposedly yielding a price sufficient to compensate for the risk of the option being exercised against them). On the other hand, if oil is only at $80 in December 2012, the option will be "in the money", and the option buyer will certainly be seeking to exercise their pre-purchased right to sell at $100.

To whom will they be selling? As with futures, the proximate counterparty for these exchange-traded derivatives is always the exchange itself – but again, as with futures, the option buyer has only been able to obtain this option in the first place because someone else was willing to sell it on the exchange. This someone else remains on the hook for this exposure with the exchange, and if the option they sold is in the money on the December 2012 date, they will be obligated to buy oil from the exchange at $100. In a very similar way to that seen with futures, the trade in options gives rise to a paper quantity of notional barrels at stake in the net exposure market participants build up through their dealings – and these paper barrels can be counted alongside those of straight futures in totalling open interest.

2.6 The Phantom Menace

It is the paper barrel of the futures contract that sets the price for the physical barrel in the real world. The paper barrel is more phantom than physical, because the fact of the matter is that less than 1% of the oil futures traded on Nymex actually run to physical delivery – the other 99% are instead settled financially, through the offsetting mechanism described above. Of course, even legitimate physical producers or users of a commodity nevertheless find it easier to settle futures through offsetting. If it were indeed such commercial market players accounting for the bulk of contracts, whether offset before expiry or not, we could likewise be sure that it was their view of how markets were developing, whether ultimately proved incorrect or not, that was guiding price discovery. This is important because futures markets are in the end intended to help actual physical producers and users of a commodity manage their business, rather than provide an arena for speculative profits.

Yet a 99:1 ratio of offset to physically-settled oil futures contracts on Nymex leaves an awful lot of room in which to postulate speculators – in other words, those who by the Nymex definition above do not produce or use oil – playing a major role in price discovery. It is also remarkable how the absolute *volume* of barrels notionally at stake in these contracts, each for a thousand barrels, compares to actual physical consumption of oil in the real world. When the Bank for International Settlements (BIS) looked at the oil futures markets in a 2007 report, it found that whereas in 2002 the largest contract (in volume terms) traded on the Nymex and the largest such contract traded on ICE together accounted for, on average, the barrel-equivalent of 3.2 times global daily oil production, by 2005 this ratio had grown to 3.9 times. Fast-forward to 2008, and do the same sum on the basis of data for Nymex and ICE front-month contracts supplied by Platts Oilgram. On May 22 that year, outstanding volumes totalled across the two leading front-month contracts were equivalent to some 509 million notional barrels of oil, or at least 5.9 times global daily production at that point.

It might be reasonable to suggest that between 2005 and 2008, proportionately that many more commercial producers or consumers of physical oil had realised the value of hedging and had entered a futures market they had previously played no part in. Or perhaps instead, to suggest that through this period those already in the market had decided to hedge a lot more of their production or consumption than they had done previously. And indeed, rapidly-appreciating prices could explain this latter behaviour for both producers and users, the former locking in ever-greater shares of production at ever higher prices out of greed, the latter locking in prevailing high prices out of fear that the trend means they will go higher yet. Many would argue, however, that this growth in paper barrel trade without the requirement of physical settlement was instead symptomatic of a significant growth in the speculative element in oil futures markets.

Some degree of speculation, particularly by market-makers, is essential in the provision of liquidity to a futures market. But whenever price discovery as a function becomes more determined by the judgment of speculators than by physical producers or users of a commodity, then by extension the prices that emerge in that market are not going to be seen as reasonable by at least one side of the producer or user divide. To that degree they are going to be both losing money on hedges previously put in place at what they felt were reasonable prices, and also more reluctant to use the futures market, so this market will therefore be failing in its stated purpose. This point was made by Masters when, in his testimony as above, he said, 'the problem is that because of the price volatility that has been created in these markets, many people that used to hedge are now not hedging... people are literally afraid to hedge, and we've talked to a lot of physical producers, some of which did hedge at much lower prices and have had to cover or rebuy what they sold at much higher prices, which in some cases has forced them out of business'.

The same point was also made during the course of the same hearing by Douglas Steenland, chief executive of US airline Northwest, whose written testimony stated:

> In our view, the volume of speculative activity has translated directly into speculative momentum that has placed upward pressure on oil prices irrespective of market fundamentals. The original purpose of the futures market – to allow end users and producers to hedge against future price fluctuations – has been subverted. When futures prices are driven by speculative activity, hedging becomes a high-risk activity for producers and users. Because futures prices are a determinant of spot prices, end users of petroleum products, such as airlines, are forced to pay prices which bear an uncertain relation to the economics of oil production or consumption and therefore cannot be planned for.

If it is speculators rather than physical producers or users of a commodity driving price discovery – the menace of the phantom barrels – the price emerging in these contracts will accordingly be divorced from the underlying fundamentals of the physical market as perceived by those best-placed to judge them. And the first line of any such argument alleging speculation as the main cause for the rocketing oil price invariably drew attention to precisely this. The obvious signals from physical oil markets that might support the appreciation of the oil price to those levels on the basis of supply and demand fundamentals were curiously lacking, despite the number of times people repeated the arguments relating to potential supply disruption. This line of argument is akin to the stance famously taken by Sherlock Holmes in the Arthur Conan Doyle short story *Silver Blaze*, where the great detective advises a police inspector to take note of 'the curious incident of the dog in the night-time'. As the inspector protests that the dog did nothing, Holmes remarks, 'That was the curious incident'. Which dogs were not barking as the oil market was seeing its very own smash-and-grab played out?

2.7 Dogs That Did Not Bark

As we have seen above with the Saudis, OPEC president Khelil and others, throughout the run-up in oil prices, OPEC countries had maintained the line that actual physical supply shortages were not to blame since the actual daily, physical oil market remained balanced – nobody who wanted oil was having to go without. Of course, it could be argued that this was exactly what OPEC would say to shift blame from itself while still enjoying higher prices. But Western, commercial, private sector oil companies were also saying this. UK oil giant BP hosts a launch presentation for its annual *Statistical Review of World Energy*, a document invariably picked up and quoted in the press when it emerges every year. And speaking at the 2008 launch in June that year, the company's chief economist Christof Ruehl was quite adamant there was no physical shortage of oil in the market – 'The market,' he said confidently, 'remains well supplied.'

Oil inventory levels

There is evidence to support this assertion in the most basic sense. We noted earlier that there are various disparities between even reputable sources of oil supply and demand monitoring. No one can ever verify exactly how many barrels of oil are produced worldwide each day or how many exactly are consumed, and neither should we expect them to. But happily, these partially-veiled fundamentals manifest their relative weight to one another in the far more immediately discernible shape of oil inventories, the stocks of crude oil held in storage around the world. It is these stocks that in the first instance meet excess demand over supply at any given time. And indeed there are regular, cyclical peaks in crude oil-derived product demand – such as the onset of the gasoline-intensive summer "driving season" in the US, or a higher call on diesel-fired power stations over cold winters – that cause refiners to build crude oil stocks beforehand, as a precaution against unexpected interruptions in their chain of supply locking them out of these lucrative trading windows.

These cyclical stock builds occur over longer-term patterns of stock-holding reflecting the geopolitical currents we have already analysed. In times of tension, people tend to hold higher levels of crude oil stock than in more relaxed times. Of course, as with the primary production and consumption data, there are disparities and some "guesstimation" involved in global oil stock monitoring. At the end of the day, however, barrels in storage are relatively easy to count and most countries see provision of accurate data as a public good that benefits all of them by allowing timely adjustments in production when required. At any given conjunction of cyclical and underlying factors, these reported oil inventories can inform us of the basic state of the underlying physical market. Rising stocks of oil in storage mean more oil is being produced than is required for consumption at that point in time; falling stocks of oil in storage tell us that more is required for consumption than is being produced at that point in time.

What were oil inventory levels through the first half of 2008 telling us about the fundamental balance between supply and demand? Certainly, successive quarters of apparent stock draws had occurred through 2007 and fed into the feeling that demand was running ahead of production. Things changed in early 2008, however. If we go right back to that first day of trading in 2008 when oil actually initially touched $100 per barrel, it did so as the US benchmark Dow Jones equity index was dropping due to poor US manufacturing figures. The recession – which many had feared was on the way – was on the cusp of being confirmed; and all else being equal, a slackening of demand growth should have eased the pressure on both inventories and oil prices. Indeed, this prospect was widely seen as responsible for the retreat in oil prices back towards $90 per barrel that occurred through January and February 2008. No matter that crude prices had definitely established themselves above $100 by early April, at this point the EIA was publicly forecasting that US petroleum consumption in particular would actually decline through 2008 as a result of the economic slowdown (the agency did still expect global 2008 consumption to increase overall on the basis of partial data showing continuing growth in China, India, Russia and the Middle East – although we know now with the benefit of hindsight that global consumption overall would also drop in 2008).

As we have seen, supply disruption was not just feared but had also come to pass in the first half of 2008 – particularly in Nigeria. Yet were these disruptions enough to tighten the market further, despite an emerging recession? This does not seem to have been the case. According to EIA data, in both the first and second quarters of 2008, crude oil inventories actually grew across the OECD region (essentially the developed world) – from 4.08 billion barrels at the end of December 2007 to 4.09 billion barrels at the end of March 2008, and 4.13 billion barrels by the end of June 2008. Another dataset from the IEA, the OECD's own energy agency, confirms that same picture with only slight variation in the figures. Meanwhile, Saudi oil minister Ali Al-Naimi had repeatedly commented through this period on how inventory levels through the OECD region were within the five-year average band and could therefore in no wise be seen as cause for concern in terms of physical supply security. While some initially disputed this, the IEA itself had announced in March 2008 that OECD stocks were 'almost uniformly above the seasonal norms', and had even actually been above the five-year average from the end of January onwards, a situation which would not alter through the coming year. If we are looking for fundamental support for high oil prices, among several indicators that alert us to an underlying actual supply shortage, oil inventories are the most obvious – yet this particular dog was definitely not barking. We can clearly see in oil stock data the empirical evidence for Ruehl's comment on the market being 'well supplied' in June 2008.

The industry cost consensus

If high prices therefore were not reflecting any immediate physical shortage of oil, what else in the spread of fundamental factors might justify them? We have already noted the "marginal cost of supply" argument, the idea that the price of oil has to be high enough to justify the production cost of the least economic barrel required to meet consumption. Human beings are rational creatures and we do not just live in the here and now but are continually forecasting and planning for the future. Thus high oil prices might not necessarily reflect a fundamental physical shortage of crude in the present but might instead simply be reflecting the highest-cost production seen as required in the capacity mix to ensure that demand is comfortably met. The only problem with this argument is that oil companies themselves were practically unanimous: even their highest-cost production currently operating or due to come onstream in the near future did not cost $100 per barrel to produce.

Let's go back to Christof Ruehl's own employer, UK oil giant BP. While BP refuses to publicise any precise oil price forecasts it makes, and indeed denies the utility of doing so, the figure it does choose to publicise instead is its own so-called 'long-term planning price'. This is the price level at which BP tests the viability of planned investments over many years of projected production prior to making the so-called 'final investment decision' (or FID), essentially the green light for project construction. As such, it is not itself an average oil price forecast but rather what the company sees as a viable floor for this average oil price, a level beneath which prices are unlikely to drop. And what in turn provides the confidence that a given price level should form the floor is none other than BP's own reckoning of how much it will cost to produce oil through the period in question – the planning price is derived from a marginal cost of supply-type analysis.

When BP reported its 2007 full-year results in February 2008, it used the occasion to announce a change in its long-term planning price. Chief executive Tony Hayward said that from now on, BP's long-term planning price would be $60 per barrel. Certainly this was a notable increase on the previous planning price of $40 per barrel, and it was seized upon as evidence that the oil industry believed that there had indeed been a permanent shift upwards in the oil price. But $60 is a long way from $100. What BP was saying with a long-term planning price of $60 was that the most marginal supply it currently foresaw developing in its own portfolio had a cost of production that allowed a reasonable return on long-term project investments if prices averaged $60.

Of course, it could just be that BP itself was a particularly conservative company that used a deliberately low-balled price figure for project planning, perhaps not reflecting what other large companies were getting up to. There is some truth to this, as BP has largely steered clear of the relatively high production cost "unconventional" supply sources its UK-listed rival, Shell, has made large investments in (particularly Canadian oil sands). Yet in the following months it became clear that while other oil industry professionals might entertain slightly richer ideas for where the oil price "needed" to be to reward sufficient marginal production to meet demand, BP was merely toward the bottom of a fairly limited band of estimates. The consensus among professional oilmen who voiced opinion on this was very much that the highest-cost production required to meet prevailing demand, invariably cited as either Canadian tar sands production or very deepwater and complex offshore development, was actually economically viable at oil prices significantly below $100.

We know this from on-the-record statements offered by prominent oil industry executives called before the US legislature in the hearings we have already referred to regarding high oil and related high fuel prices. In one session before the Senate on May 22 2008 were gathered together John Hofmeister of Shell's US subsidiary, Robert Malone of BP's US subsidiary and John Lowe of US major ConocoPhillips. Hofmeister said that if it reflected the industry cost base, the price of oil should be 'somewhere between $35 and $65 a barrel'. Malone actually mirrored his company's planning price in saying 'in the range of $60'. And Lowe went the highest, arguing that the price should be 'about $90 in this environment'. These opinions were being offered on a day when oil actually passed $130 per barrel on Nymex for the first time ever, to close at a then-record $133.82.

Looking at this evidence, it remains very hard to justify oil prices over $100. Neither current physical supply deficits rationalise it, nor the production costs of marginal supplies expected as necessary to meet demand any time in the foreseeable future. The obvious alternative for many observers inside the oil industry was that it was indeed the activity of speculators in the futures market, bidding up prices beyond levels justified by these physical fundamentals. It was not only oil company executives called to the long-running series of US legislative hearings on the oil price – professional oil industry consultants and analysts (as opposed to investment bank analysts) were also there in force too. One of these characters, Fadel Gheit of Oppenheimer & Co, had already made headlines in December 2007 when at that point, before oil had crossed the $100 threshold, he testified before a Congressional committee that speculators were responsible for oil costing twice as much as it should, even when it was around the $90 per barrel range.

Six months later, on June 23 2008 – and with the oil price by then tracking between $130 and $140 per barrel – Gheit was once again testifying in front of a Congressional committee on the oil price. The following is extracted from his written submission accompanying that day's testimony, and neatly summarises the points above, that when physical market fundamentals fail to explain oil price movement, the obvious cause is instead speculation:

> I do not believe the current record crude oil price is justified by market fundamentals of supply and demand. I believe the surge in crude oil price, which more than doubled in the last 12 months, was mainly due to excessive speculation and not due to an unexpected shift in market fundamentals. After all, demand growth in China, India and other developing countries was not a surprise and was reflected in crude oil futures a year ago. In fact, the slowdown in global economic growth, caused by the sub-prime debacle

and fears of a run on banks, trimmed world oil demand forecasts, which should have resulted in a lower, not higher, oil price. On the supply side, the impact of the unrest in Nigeria on oil exports, the decline in Mexico's crude production, and other less than newsworthy factors, were hardly new news, and were already discounted in crude oil futures. I firmly believe that the current record oil price in excess of $135 per barrel is inflated. I believe, based on supply and demand fundamentals, crude oil prices should not be above $60 per barrel. My view is based on the following observations: there were no unexpected changes in industry fundamentals in the last 12 months, when crude oil prices were below $65 per barrel. I cannot think of any reason that explains the run-up in crude oil price, besides excessive speculation; world oil demand growth forecasts have been trimmed to reflect the current economic slowdown, which should have resulted in lower, not higher, oil prices; world oil production is economic at prices well below $65 per barrel.

Testifying alongside Gheit that day were not just hedge fund manager Michael Masters and fuel-consuming industry representatives such as airline chief Douglas Steenland, as noted above, but also another two professional oil industry consultants, Roger Diwan of PFC Energy and Edward Krapels of Energy Security Analysis. Strikingly, all of these witnesses would essentially agree with the angle offered by Gheit. At one moment during the oral testimony from this panel of witnesses on June 23 2008, one Congressman Barton on the committee asked them collectively what they thought the price of oil should be at that moment. Back came the answers:

Gheit: 'Between $46 and $60, no more than $60.'

Diwan: 'Below $100.'

Masters: 'Below $100.'

Krapels: 'I'm in the $60 camp.'

To understand how these industry insiders all reached the same conclusion that something other than fundamentals was driving prevailing oil prices, we have to understand what was widely acknowledged to have happened in the paper oil markets in recent years. This is commonly referred to as the "financialisation" of oil, part of a wider financialisation of commodity markets in general.

3

The Financialisation of Oil

'Most frustrating for oil analysts is the lack of transparent data to look at the exact size and flows of the different categories of players on the oil markets. The data released by the CFTC does not allow analysts like me to understand the broad movements by the financial players in the market. The categories used by the CFTC are too broad and are meaningless if you allow the index funds to be categorized as commercials and if you don't break down these categories in smaller and better defined entities. I find it ironical that we constantly require more data transparency from the OPEC producers, be it on their production or reserve data, but we cannot manage to do the same in our own backyard on one of the most crucial if not the key determinant of oil price formation.'

Roger Diwan, PFC Energy – testimony submitted to US Congress, House of Representatives Subcommittee of Oversight and Investigation, Committee on Energy and Commerce, June 23 2008

3.1 The Hearings

It should come as no surprise that in the US, at one and the same time the pre-eminent global military and economic power yet also completely dependent on imported oil and a society built around the car, the rapid ascent of gasoline prices as a result of crude oil appreciation had been met with public concern. And naturally, this concern quickly turned to outrage at the suggestion that such prices might be artificially elevated through the activity of speculative finance. This translated into investigation by the US Congress itself. Hearings on the matter had proceeded at various levels of Senate and House panels throughout this period, from the autumn of 2007 into December, then March and April of 2008, and again through May-July 2008. And by the time the latter wave of hearings rolled around, oil prices had rocketed so high that public attention was keenly focused on the testimony.

It is not my intention to bludgeon the reader with extensive cataloguing of dates, the elected US senators and congressmen sitting on whichever subcommittee and their opening statements, nor the detail in the proffered complaints from associations representing airlines, truckers and others drawn from the rich panoply of American life suffering badly with sky-high oil prices. Suffice to say, these hearings largely revolved around the same issue, now commonly referred to as the "financialisation" of oil. This is the idea that the weight of non-commercial, by definition "speculative", interest in oil futures trading on the part of financial institutions had altered the dynamics of price discovery in that market so much that oil prices now reflected the imperatives of finance more than the underlying fundamentals of physical supply and demand. So some of the evidence presented by some witnesses at these hearings provides a useful prism through which to examine and weigh the arguments for and against the role of speculators in driving high oil prices.

As we have seen, throughout these hearings the US legislature invited various expert witnesses from the energy industry to offer their evidence regarding this issue, including professional oilmen and also specialist oil sector consultants. Many – but not all – of these emerged as what we might call witnesses for the prosecution. In other words they agreed with the notional charge that this financialisation of the oil futures market was introducing undue speculation which was now indeed driving oil prices higher than fundamentals warranted. And from the welter of government agency initials thrown up in these hearings, as the legislature called them to address these complaints and offer their own evidence on the matter, one set really stands out as continually arguing the defence against this charge. The GAO (Government Accountability Office) was called, the DoE (Department of Energy) was called, as too was its subsidiary EIA (Energy Information Administration, as above), along with the FERC (Federal Energy Regulatory Commission). But it was the CFTC – or the Commodity Futures Trading Commission – which persistently asserted that speculative futures trading had no significant impact on the price of oil.

Evidence from the CFTC was crucial to all these hearings because, as the US government body responsible for regulating commodity futures markets in the US, only it possessed the detailed statistical data that could be reviewed to determine if movement in speculative capital was linked to movement in oil prices. Other government agencies deferred to its judgment – when EIA administrator Guy Caruso was quizzed by the Senate in December 2007, he conceded that he personally thought that speculative investment in oil futures had perhaps contributed something to prices, but said such financial sector interest was not the overwhelming driver behind price movement because the CFTC had apparently established this was not the case. Yet CFTC evidence in this regard would prove controversial – and not only because any assertion that the oil price was being dictated by financial speculators was a serious criticism of the CFTC itself.

After all, the CFTC's main role is to ensure that the futures markets are able to perform their primary function of hedging prices for commercial producers and consumers of the commodity in question. It should, then, notionally prevent precisely the sort of "artificial" skewing of market prices by non-commercial interests that many of the expert witnesses alleged was occurring. Accordingly, it had to suffer brickbats from some irate US representatives convinced it had been asleep on the job. Beyond the obvious self-interest at stake in the CFTC assertion that speculative finance was not calling the tune in oil prices, however, many outside experts have a fundamental problem with the way the CFTC presents its data on Nymex market participants, and said so in their evidence. To understand what this problem is, we need to understand who the players are in the financialisation of oil, how the CFTC monitors their relative importance in the market, and how other commentators think the CFTC has got it all wrong.

3.2 The Meaning of "Financialisation"

"Financialisation" is a neologism now frequently used to describe what has happened to the crude oil market over the past few years. In its basic understanding, the word simply denotes increasing involvement in that market by a whole fauna of financial investors – from lean, mean speculating machines (such as hedge funds out to make a quick killing) to large, lumbering beasts (such as traditionally staid and risk-averse pension funds just looking for some diversity in their required diet of stable, long-term returns). We can refer to this whole range of players as "institutional investors", as they all represent investing institutions of one sort or another. What unites them apart from this definition is that while they are keen to wring some form of financial return from their foray into the sector, they have no interest as businesses in the actual physical production, supply and end-usage of crude oil.

This is in stark contrast to the oil producing companies, oil traders, refineries, fuel marketers, and intensive fuel-users such as airlines, who were for years the main protagonists in the oil markets, and whose views on the fundamental "real world" factors affecting the supply and demand of actual physical oil were traditionally seen as determining the oil price. Which brings us to other meanings suggested in "financialisation" – referring not just to the simple fact of increasing institutional investor interest in the oil markets, but also to how the weight and momentum of this financial interest has allegedly become the dominant determinant of oil prices, obscuring "real world" fundamentals. Thus the marketplace behaviour of oil gradually shifts from that of an actual physical good – with, for example, price-setting demand following predictable "seasons" in terms of annual patterns of peak and trough physical oil usage – towards that of a particular class of financial investment asset, with price-setting demand set by the characteristics of the return or "yield" on offer in comparison to other investment asset classes.

There is clearly a sliding scale of meaning in the usage of "financialisation", and some definitions comfortably nest within others. For example, David Kirsch of oil consultancy PFC Energy was apparently using the term in its simplest sense when interviewed by Arab News at the very start of 2008. Following the first $100 per barrel trade he explained the 'financialisation of oil' simply as the use of oil as an investment product by hedge funds and pension funds. Yet some months later his PFC Energy colleague Roger Diwan was using the same term in a far more complicated sense, when testifying to the US Congress in the June 23 2008 hearings. In fact, he came up with a threefold evolution of financialisation in the oil markets over the previous few years. This had started with a phase running from around 2003 to 2006, through which financial investors simply bought into oil futures en masse, all hoping for positive price appreciation in their investment based on the tightening fundamentals we have sketched in our first chapter – and indeed, many were rewarded for buying into this story. Nevertheless, according to Diwan, from 2006 onwards a second stage of financialisation evidenced far more measured buying and selling by financial institutions under more sophisticated strategies – involving going short of the oil price at times, as well as long at others.

For Diwan, the third stage of financialisation was evident from the onset of the credit crunch in August 2007, and saw a fresh concentration of non-commercial, financial sector interests in the paper oil futures market. This was made up both of money fleeing other types of asset classes which had become distressed in the mass deleveraging and sell-offs attending the credit crunch, and also new money born of the deep interest rate cuts governments adopted to ease the crunch. This latest stage of financialisation was causing oil futures to respond as much to dollar weakness in the currency markets and perceived inflation dangers (variables that affect financial sector demand for oil as a perceived "safe haven" hedge against dollar debasement and inflation) as to the variables reflecting physical supply and demand fundamentals.

Diwan was of course not the first to point out that an inverse correlation had become apparent between the oil price and the strength of the dollar or expectations for inflation – which are, of course, aspects of the same phenomenon for holders of dollar-denominated funds. This relationship was widely commented on throughout the media, and indeed was also part of the standard OPEC line. As often as cartel representatives mentioned speculators, they also mentioned that it was gyrations in dollar strength motivating the speculators in their oil investment strategies, as much as their long-standing appreciation of the potential for supply disruptions. In early June '08, for example, OPEC president Khelil had commented: 'In terms of market fundamentals, there is no problem with supply and demand. Prices have more to do with this speculative bubble, which is due to the depreciation of the value of the dollar and geopolitical tensions.'

This linkage is in fact a key part of the investment case encouraging financial institutions into commodities in the first place. Producers of dollar-denominated commodities such as oil are expected to raise the dollar price of their product if the dollar weakens against their home currency – so that the value of their sales are maintained once repatriated – and likewise raise the dollar price of their product in line with dollar economy inflation. Thus, investing in dollar-denominated commodities provides a "natural hedge" for investors against the twinned evils of dollar devaluation and inflation. Of course, for most of our modern history, it has been dollar-denominated gold that has reigned supreme as the commodity "safe haven", but over recent years oil and other metals (certainly platinum) have also been favoured. And Diwan's own statistical analysis presented as part of his testimony demonstrated that since August 2007, while the so-called "R-squared" measurement of statistical correlation between strength in the gold price and weakness in the dollar was 78.8%, the R-squared between the oil price and dollar was almost as high, 75.9% – which prompted Diwan to label his graph showing this relationship as 'Oil the New Gold'.

3.3 The Growth in Speculative Interest

Taking the examples given above as a guide, we can distinguish between a "soft" and a "hard" version of oil market financialisation. The former simply refers to rising financial sector interest in investing in oil; the latter instead positing some kind of distortion of the market, such that this financial sector interest and its own concerns for maximising yield on its managed money have become as key a determinant of oil prices as the physical fundamentals of the market. Looking at what statistics there are, it would be hard for anyone to disagree with the "soft" meaning of financialisation at least. As Diwan noted, between 2003 and 2008 the number of financial sector companies directly trading on

Nymex grew from 50 to more than 250, and this figure does not include the financial sector players investing in the market through intermediaries, such as the "swap dealers" we will examine below.

Figure 6 is drawn from CFTC data, and presents snapshots of open interest across all Nymex light, sweet crude oil contracts on three different dates, split into the two categories the CFTC uses in monitoring all market participants, so-called "commercial" and "non-commercial" traders. The commercial category includes all the real-world producers, traders and users of physical crude oil (such as those listed above) who directly maintain positions on the Nymex futures market to hedge their existing real-world price risk. The non-commercial includes all the "speculative" market participants whose interest in oil futures is purely financial rather than anything to do with managing real-world business risk. Historically, this split between commercial and non-commercial trading activity using these definitions has been the CFTC's preferred format for reporting weekly market data to the public in its so-called Commitment of Traders (CoT) report – so, as we shall see, historically this has also been the best data available to external analysts (barring some very recent exceptions, also detailed later).

We have already identified one such "speculative" class of trader, which in fact the futures market would be hard-pressed to do without – the market makers. Alongside them the non-commercial category includes "floor brokers" (those market participants trading futures on the Nymex floor for either their own financial gain, or as a service to external clients lacking their own market access), and also the hedge funds that seek to invest directly in oil futures. All these "non-commercial" parties are *a priori* assumed to be speculative traders with no actual business interest in the physical supply and consumption of crude oil. They invest in exactly the same contracts as physical market players but with different aims, generally seeking to simply realise gains on offsetting, rather than pursuing any actual physical hedging requirement (although hedge funds and other financial institutions do, occasionally, extend their interest in oil into the physical market).

One trade in particular is seen as characteristic of the "non-commercial" market participants – the so-called "time spread" trade. Time spread trades involve a speculator going long of the contract price at one calendar point along the futures curve, and short of the contract price at another point of the futures curve. Their motivation is that they feel, for whatever reason, that prices at one part of the curve are out of whack with prices at another, and they expect the differential between them to either narrow or widen. They may not, however, be sure how the prices will move to achieve this outcome. While one price will be too high or too low with regard to the other, will it be this price that moves to narrow or widen the gap, or the other price, or both prices moving at once? Providing the time spread speculators go long of the price they think is too low relative to the other, and short of the price they think is too high relative to the other – as long as they are correct about whether the gap in question will narrow or expand – they will make money.

Much as with a "pairs trade" in equity markets, even if both prices move in the same direction but the trader is still right about the *relative* performance of the two prices, then losses on the price that has moved in the "wrong" direction relative to the trader position will be outweighed by the gains in the other price moving in the same direction, but to a greater extent. As with an equity market "pairs trade", wrongly predicting whether or not the gap will narrow or widen is the one way such a trade can go wrong, but it may well then go disastrously wrong – as the trade will possibly be taking losses on both positions. Time spread trades going wrong can therefore occasionally be responsible for notable breakouts of simultaneous buying and selling at different points along the curve, as speculators dash to reverse the positions. One such example arguably occurred during a particularly fevered week in May 2008, which we shall examine later.

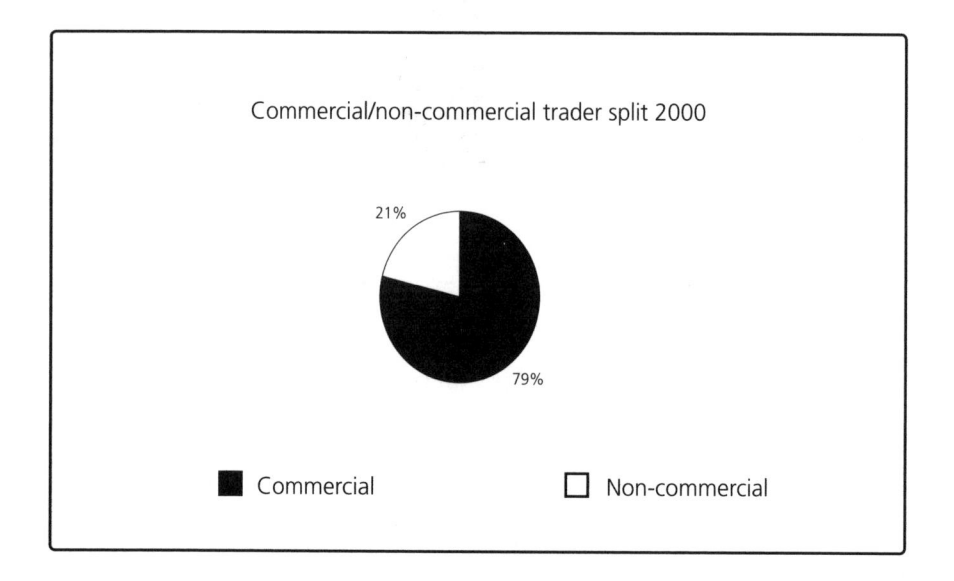

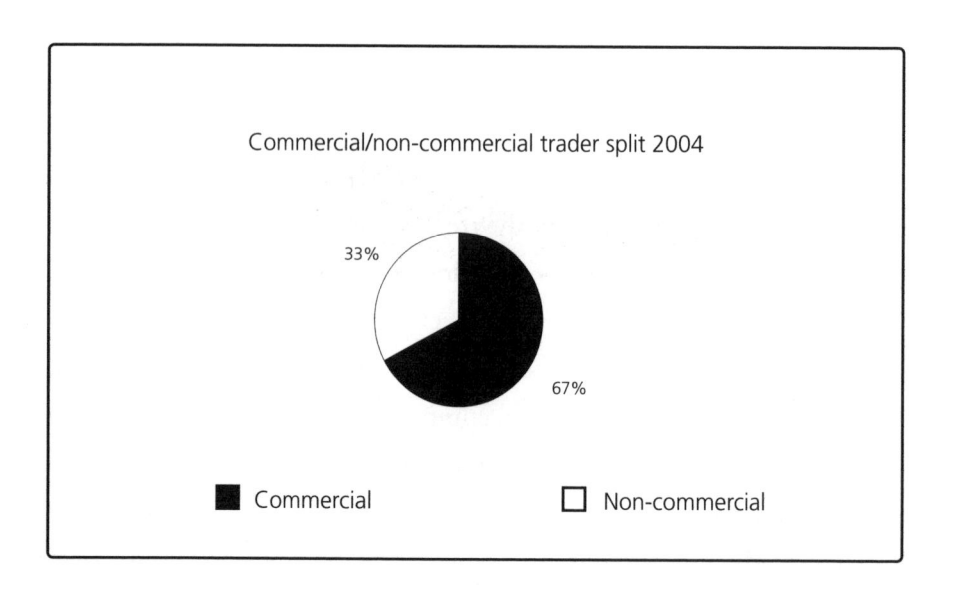

Commercial/non-commercial trader split 2000

- Commercial: 79%
- Non-commercial: 21%

Commercial/non-commercial trader split 2004

- Commercial: 67%
- Non-commercial: 33%

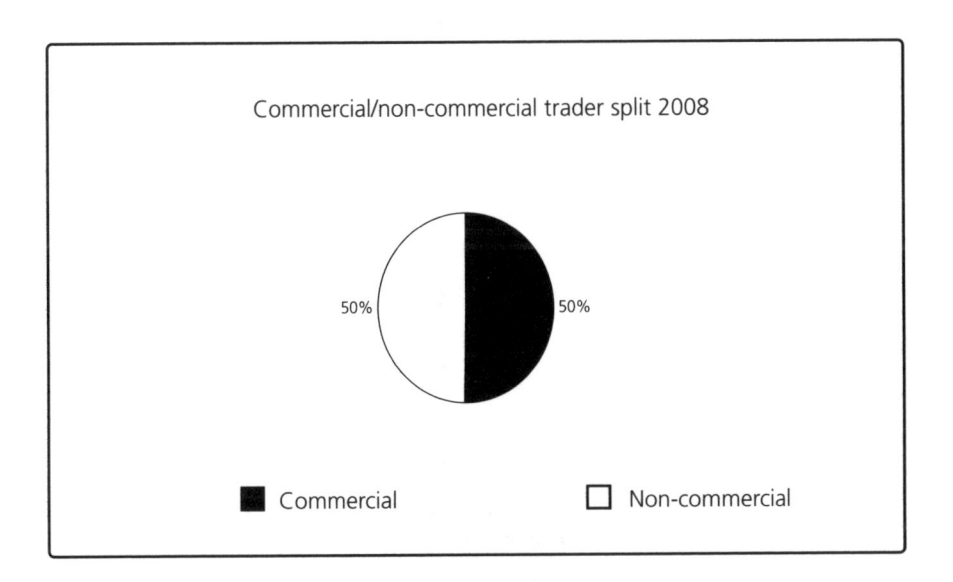

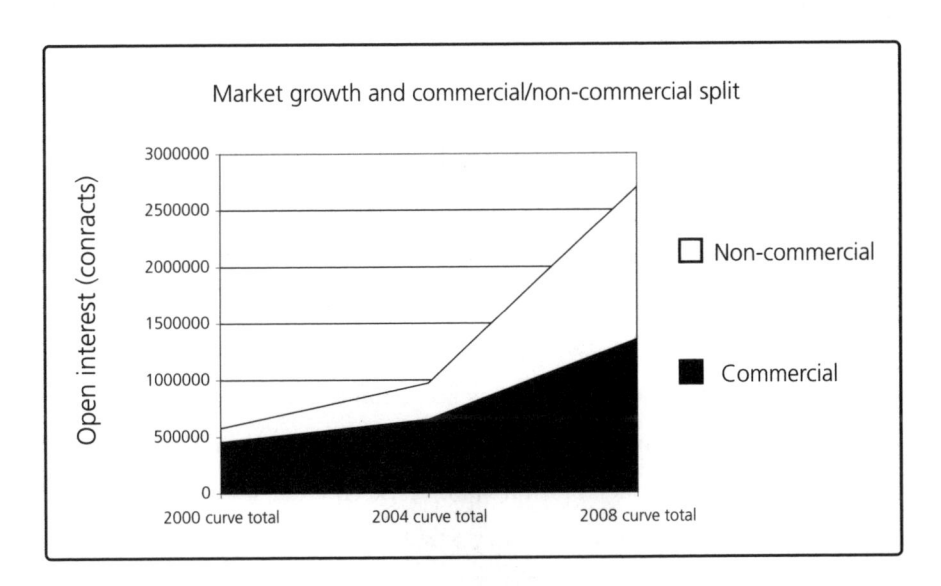

Figure 6: Nymex oil market growth by CFTC "commercial"/"non-commercial" split [Source: CFTC (Büyüksahin, Haigh, Harris, Overdahl & Robe, 2008)]

As can be seen in Figure 6, while in 2000 the non-commercial side of the market accounted for 21% of total open interest, by 2004 this had grown to 33%, and by 2008 to 50%. Anyone tracking this data saw undeniable growth in the non-commercial, speculative side of the futures market. Yet the CFTC was adamant that the growth of non-commercial trading was not itself a driver for high oil prices. How did it establish this? Let's go back to that CFTC finding relied upon by EIA administrator Caruso. This was an important document as far as the CFTC was concerned, and was also referred to by CFTC Commissioner Walt Lukken himself when commenting on speculative pressure on oil prices in December 2007 (before the Congressional Oversight and Investigations Committee on Energy and Commerce).

It is worth quoting Lukken at length with regard to the questions that study addressed and what the findings were:

> Recently, the CFTC's Office of Chief Economist examined the markets and the role that speculators play in them. The staff studied the relationship between futures prices and the positions of managed money traders (MMTs), commonly known as hedge funds, for the natural gas and crude oil futures markets. The staff also examined the relationship between the positions of large speculators such as hedge funds and positions of other categories of traders (e.g. floor traders, merchants, manufacturers, commercial banks, dealers) for the same markets... The study found that when new information comes to the market and creates some price movement, it is the commercial traders (such as oil companies, utilities, airlines) who react to it first... Hence, the report's conclusions show that speculative buying, as a whole, does not appear to drive prices up.

Despite the comment that the work was recent, this study in fact originally appeared as a working paper in 2005, two versions of this still being easily available – and was then apparently updated in June 2006, although I have not seen the later version.

Hedge funds are of course the very definition of speculative capital, and as such have become the *bête noire* in almost any modern tale of financial excess, from the emerging markets collapse of 1997-8 onwards through to the sub-prime mortgage credit disaster that triggered the current global recession. It is easy to see why the CFTC would focus on hedge fund trading in an analysis of speculative pressure on the price of oil futures. After all, it would be unfair to treat all non-commercial participants as sources of speculative pressure – as we have seen, market-makers are technically counted as non-commercial, yet their presence is essential in providing the liquidity needed for an efficient market. And while some floor brokers, such as Richard Arens, were indeed speculators out to play the market, others were simply service providers for off-market clients of whom many would actually have a genuine business interest in physical oil but lack the scale or desire to trade on Nymex themselves. As hedge funds are presumed to be solely interested in the financial gain to be won on correctly anticipating oil futures price movements, they did indeed make a suitable subject for examining if such speculation had an observable correlation with movements in the oil price.

This was moreover a study only the CFTC itself could have undertaken. Only the CFTC had access to more detailed trading records than the simple split between commercial and non-commercial traders it presents to the world in the CoT reports. Indeed, in the original working paper the authors make note of this consideration themselves in vaunting the "unique" nature of the dataset to which they had access, a dataset which split the two master categories down into more particular subsets, including hedge funds. The conclusion drawn was that there was no significant correlation between the investment patterns in Nymex oil futures by hedge funds and the movement of the oil price itself. This study was the initial basis of the stand the CFTC took right from the start of the hearings – that despite the obvious growth in non-commercial interest on the Nymex crude oil market, the speculative element of this interest was not responsible for driving oil prices so high.

3.4 Enter the Swap Dealers

Outside commentators such as Gheit, Diwan, Masters, and Krapels, disagreed with this conclusion. In voicing their objections they raised a major issue with the way the CFTC presents its data, and how this influences any attempt to gauge the true effect of speculation on oil prices. The problem lay in the initial categorisation of market participants into commercial and non-commercial. A particular type of market participant included in the commercial categorisation, and therefore of necessity excluded from the CFTC definition of "speculative" adopted to date, should arguably have been more accurately classified as non-commercial, and therefore part of the speculative interest. These market participants are called "swap dealers", and the perceived "back door" through which this supposedly commercial category of market participant was and is seen to exert considerable speculative pressure is commonly referred to as the "swap dealer loophole".

Sometimes a business or individual will have such exacting, particular requirements in how they wish to fix their price exposure along some portion of the oil futures curve, whether front-month or further maturities, that simply buying and selling standardised futures contracts on Nymex or similar markets does not do the job. They will have to turn instead to the so-called "over-the-counter" (OTC) world of bespoke, tailored derivatives brokered privately rather than on regulated markets. Such derivatives are in general proffered by financial institutions who reckon themselves capable of accurately calculating the risks they expose themselves to in guaranteeing to pay out – under a 'swap' agreement – the difference between a price fixed now, and a possible future price for a commodity (the conditions for payout of course depending on whether the customer wants to go long or short of the strike price specified and in which timeframe).

Often the swap dealers are investment banks, able to deploy legions of mathematical analysts and terabytes of computer processing power to calculate the optimum prices they should charge for taking on this risk while winning the largest possible number of customers. In this business model, the ideal position for the swap dealer is to be completely hedged itself, in other words neutral to the financial outcomes of the derivatives it has written whatever happens to price levels, while still raking in fees from supposedly taking on price risk. And of course, the business of writing swaps itself provides a natural hedge in this regard. Theoretically, in many markets there may be as many people wanting to go long of a particular price at a certain level as there will be wanting to go short at that same price. If the swap dealer can write the derivatives for all these customers it will gain with as many as it loses, and to the same extent on either side – no matter what happens to the price in question. As the opposing counterparties are therefore effectively funding each other's potential gains, the swap dealer is left in the middle – collecting fees while netting off these opposite exposures against each other, putting together different ends of the same trade from all suitably opposing positions on its book.

More often than not, of course, in the real world there will be an imbalance between the number of customers the particular swap dealer has wishing, in aggregate, to go long or short of a particular price along the futures curve, and therefore it is left with a net exposure at that point. This is where the swap dealer's own trading on Nymex or whichever other appropriate futures market comes in. Say a swap dealer has written derivatives equivalent to a thousand contracts-worth of oil volumes for customers going long of five-year oil at $150 per barrel, but also written the equivalent of two thousand contracts for customers going short of five-year oil at the same price. A thousand of each of these opposing exposures cancel each other out; however, the swap dealer still has a net exposure to a remaining one thousand customers who have gone short of oil. Accordingly, the swap dealer's own exposure is financially equivalent to going long the equivalent amount – by writing these derivatives he has become effectively a net buyer of five-year oil at $150.

This is easily seen to be the case: the swap dealer "wins" on this exposure if the oil price rises, because then he will not be paying out on the short derivative he sold but these customers will instead be paying out to him. It is of course possible that the swap dealer might wish to keep this exposure, particularly if its own opinion is that the five-year oil price is indeed going to rise. However, good business practice is to hedge out this remaining net exposure, and to do so the swap dealer simply has to cover his net long exposure by selling the equivalent amount of contracts for five-year oil at $150 on Nymex, which is of course also the same as simply mirroring the net exposure his own customers hold in the first place. Now the swap dealer is completely risk-free. He will no longer benefit if the oil price rises, as the gains he makes from his OTC swap counterparty he will have to pass on to clear his own equivalent loss on the short exposure he has on the regulated market. However, likewise he no longer loses if the oil price drops, as the payout he will have to make to his OTC swap customers is funded by the gains he makes on the short position on Nymex.

The Financialisation of Oil

If the whole swap dealing business as described above sounds like some kind of clever investment bank game where they win whatever happens, bear in mind that this is exactly what the wider market expects them to do in this position. Why should they lose out for providing a very useful service often called upon by other segments of the market? The whole business is entered into by all in the full acknowledgement that the swap dealer itself will hedge its way out of most positions, and indeed it is only because of its configuration and experience as an investment bank – which already suits it to be a market aggregator, capable of pulling a single measurable net exposure from countless single exposures – that it is able to offer this valuable service in the first place.

This is the specialisation that justifies the fee payments worked into every transaction it undertakes, these fees being the sole revenue from the business if all exposure is systematically hedged away. Moreover, the business is conducted in a way that allows swap dealers to function without having to use whole slugs of their own capital as collateral when hedging their own exposure. Just as a regulated futures exchange demands margin accounts and constant collateralisation of potential losses, so too do OTC derivative customers fund margin accounts with their swap dealers and collateralise these as appropriate – allowing the swap dealer to effectively use this posted collateral as its own collateral when it needs to go to market itself.

Swap dealers have always been classified by the CFTC within its commercial categorisation, because for years such structured OTC derivatives were indeed used by commercial players in the energy industry. Moreover, the investment bank swap dealers themselves, in many cases, operate through subsidiary entities that were historically oil trading companies involved in Nymex as commercial players already. For instance, Goldman Sachs carries out all its business on the Nymex

105

exchange through J. Aron & Co, an oil trader it bought up several years ago, while Citi likewise operates through oil trading firm Phibro. Often, these same entities are involved in genuinely commercial physical oil trading alongside facilitating the hedging of their parents' financial sector swap dealing exposures, all of which can nevertheless require trading in oil futures. So it can be hard to see where the actually non-commercial market activity and the other, more genuinely commercial activity undertaken by these supposedly commercial market participants shade into one another.

The CFTC also extends special consideration to swap dealers in exempting them from the position limits that apply to all other futures market participants. Part of the CFTC armoury in supervising markets under its jurisdiction is imposed limits on the size of the positions any one market participant can take, including non-commercial players such as hedge funds and floor brokers. This is designed to prevent speculation taking over from the actual requirements of the physical market as the main driver in pricing. Swap dealers are, however, exempt from such limits. The CFTC recognises both their need to hedge away exposure resulting from their derivative-writing activity as a genuinely commercial requirement, and also that the volumes which might require hedging could be far larger than the position limit that would apply to the swap dealer if considered as a single participant under the normal rules.

As far as many were concerned, however, swap dealer involvement in the futures trading market had long since ceased to be dominated by commercial concerns, and was instead a prime conduit for speculative financial interest in the futures market. To understand why this should be the case, we need to realise that as far as speculative investment by financial institutions in oil is concerned, what was formally ascribed to such speculative interests operating on the Nymex oil futures market under the CFTC definition of "non-commercial" trading was just the tip of a very large iceberg. We need to understand the wider world of commodity-related investment, and the rise of the commodity index funds.

3.5 Commodities as an Asset Class

The emergence of commodities as a widely-followed investible asset class is well documented, and indeed whole volumes are written about the subject (a good and fairly-up-to-date place to start is William Hubard's *The Financialisation of Commodities*, a Thomson Reuters report released in April 2008). Suffice to say, many institutional investors across a wide spectrum of pension funds, investment funds, hedge funds and even university endowment and sovereign wealth funds now believe it is appropriate to hold a portion of their asset portfolio in commodities. The benefits are seen to include portfolio diversification into an asset class that arguably displays low correlation with others such as equities and, as we have touched upon, inflation-hedging, as commodity prices commonly rise in tandem with inflation and are indeed often an input into inflation.

While commodity investing has been around since the 1970s, it has only been since the start of this decade that it has really taken off with institutional investors. The decision in 2000 by Dutch state sector pension fund PGGM, the third-largest pension fund in Europe, to invest a significant chunk in commodities is seen as a watershed. Hubard reports this investment, predominantly in energy-related products, as having grown to €4.8 billion or around 5.5% of its total assets by the end of 2007. Likewise the 2006 decision by California public sector pension fund CalPERS, the largest public pension fund in the US, to invest $500 million across a spread of commodities was also widely-reported and commented upon.

How are such institutions structuring their investment in commodities? A range of options are exploited:

- **Hedge funds:** Money can simply be invested in a commodity-focused hedge fund and the manager left to get on with things.

- **Commodity indices:** Institutional investors can access products tailored to mimic the notional return on a benchmark commodity index, an artificial construction invented for the purpose of tracking commodity price performance.

- **Structured investment products:** Institutional investors can buy specially-tailored notes which promise to pay out a percentage relating to the performance of particular commodities or indeed commodity indices over a certain time-span, sometimes while guaranteeing to protect capital as long as certain trigger events do not occur (such as a breach of a specified price range).

- **Exchange-traded funds (ETFs):** Investors can now access notes traded on regulated stock exchanges which track the price of both individual commodities and baskets of commodities, with refinements extending to short or leveraged exposure to this commodity price movement.

The growth in these kinds of products, particularly commodity index-tracking investment, has been incredibly strong, although it is very difficult to get a handle on the exact sums involved. But the really important point to realise is that, as we shall examine in greater detail, the latter three options all necessarily involve some kind of trading in off-exchange OTC commodity derivatives (while the first option may also in fact turn out to do so as well). The Bank for International Settlements (BIS), the nearest thing to a global banking authority, reckoned that the notional value of outstanding OTC commodity derivative contracts had reached $6.4 trillion by mid-2006.

3.6 Constructing a Commodity Return

To see how commodity index trackers, structured notes and ETFs require underpinning by OTC commodity derivatives, and are therefore key contributors to this amorphous total, we need to examine how their return is generated. We will do so taking a commodity index tracker as our model, as the principles hold true across the other asset classes as well. And doing so will allow us to also examine the rise of the commodity index fund.

How can an institutional investor gain from what he expects to be upward movement in the oil price in the coming years? It is not as easy as it sounds. Ideally, the investor might buy some oil in the here-and-now, and simply store it for as long as they wished. However, this raises problems immediately – the oil trading and storage business has its own hurdles and barriers to entry, and besides, as we have seen, investment appetite for oil can potentially far exceed the amount of physical oil available at any one point in time. So there simply would not be enough oil to go around if everyone thought like this, or were indeed even permitted to embark on such a course by the relevant regulators – an unlikely outcome in a world of scarce oil.

Of course, as we have also seen, it is possible to benefit from movements in the oil price by investing in futures markets – but this is not so straightforward either. As noted, the institutional investor does not or indeed cannot hold a spot barrel in storage, yet any futures contract they have bought will indeed eventually mature into a spot, physical, barrel unless it is offset before maturity. Meanwhile the desired horizon for investment might be much longer than even the longest maturity available in the futures market, and there is also the consideration that dealing in such long-dated futures is also traditionally less liquid, with the contracts therefore less desirable assets.

It is possible, however, to structure a virtually permanent invested position at any point along the oil futures curve, the strip of prices running from nearby maturities out to the longest-dated contract outstanding, by "rolling" in and out of different futures contracts as time passes. For instance, if it is the front-month contract that is widely followed as the global oil price which the investor wants to track, then they might buy two-month contracts, hold them until they mature into one-month contracts, sell out of these prior to the expiry of trading, use these proceeds to buy another lot of two-month contracts and so on – literally ad infinitum, as theoretically the investor could keep doing this for as long as he has the money to do so.

Thus the investor has constructed an artificially permanent invested position at the front end of the futures curve by rolling in and out of futures contracts as appropriate. It is immediately apparent, of course, that this is not literally the same as exposure to the pure front-month price, as at some points in time the investor will not always be 100% invested in the front-month contract alone even given the "smoothing" many such rolled positions achieve by splitting their investment between the relevant months to begin with. But given the proximity of the contracts involved, it will be the closest thing achievable as a *genuinely investible* solution to the problem of maintaining a permanent front-month position.

The practice of rolling futures contracts in this fashion introduces, however, another element of risk or return into the equation, on top of the front-month price movement the investor is hoping to capture. This is the "roll yield" or "roll return", and it emerges as a function of the shape of the futures curve at any particular point in time. What do we mean by this? Futures curves have their own shape, determined by the prevailing market view at any one point in time regarding the outlook for prices. For example, an ongoing supply disruption may be causing high prices at the front end of the curve, but the fact that this is expected to be resolved within a matter of months means prices further out along the curve, perhaps at one year or five years, are significantly lower.

Alternatively, there may be ample supply and indeed growing inventories in the here and now, but concerns over the longer-term outlook for supply mean the three-year price may be higher than the front-month. When a futures curve displays prices higher at the front end of the curve than further out along the maturities, it is said to be in "backwardation", and when prices climb higher as you go further out along the curve then it is said to be in "contango". Often the two conditions can exist alongside each other in the same curve – with a prevailing contango from front-month out, to a certain maturity out along the curve, and then backwardation thereafter, in which case the curve displays a characteristic "hump"; or vice versa, with prices falling away to a certain maturity and then climbing again thereafter.

Whether a contango or backwardation is prevailing between the particular points in the futures curve through which an investor is rolling their position determines whether or not the roll yield they receive is positive or negative (see Figure 7). If the curve is in backwardation, then the yield on each roll will be positive – the investor bought two-month contracts, has held them until they matured into one-month contracts, and sold out of them at a price in excess of that which he needs to buy the next same quantity of two-month contracts. The difference is his roll yield, which can build up over time into a significant amount on top of the implied gain or loss in the actual movement in the underlying prices his originally invested capital is tracking. Likewise if the curve is in contango, each time the investor rolls his position he is receiving less for the front-month contract than he originally invested in the two-month contract – and, all else being equal, this will not be enough to buy as many contracts as were bought originally, so the investment has suffered a reduction. If the contango persists for a long time, the negative roll yield can take quite a chunk out of an investment regardless of what the underlying price is doing.

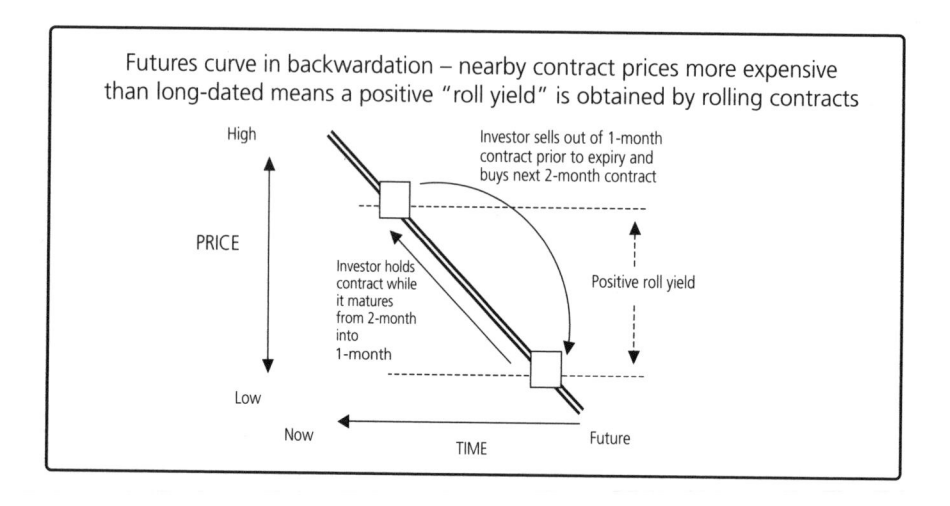

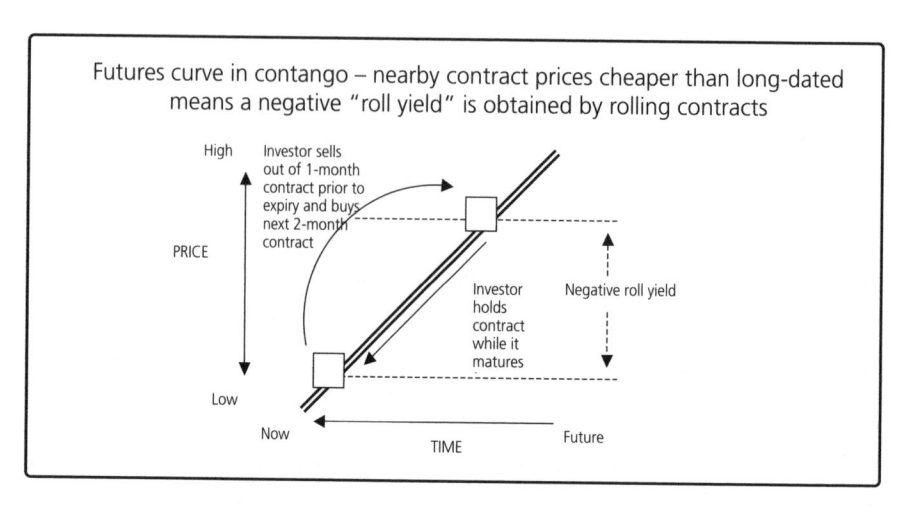

Figure 7: Roll yields on futures curves in backwardation and contango

There is a third element of return to consider in analysing any such permanently-maintained position along the futures curve. This is the "collateral yield", which simply reflects the fact that money held in margin or collateral accounts to back investment in commodity futures itself earns a standard rate of return in line with short-term deposits. This is typically seen as around the same rate as the 90-day US Treasury Bill, the "90-day T-bill rate". While all deposits held on margin for futures trades earn such rates regardless of whether they are part of a rolled position, the collateral yield becomes an important if small positive element of return when rolling a position means a significant amount of capital remains tied-up in margin or collateral through the life of the investment.

The three elements above, the so-called "spot" return (or loss) on the movement in the underlying price being tracked, the roll return consequent on holding the position as time proceeds, and the collateral return on the capital tied-up in these trades, characterise the return profile for maintaining such an invested position at any particular point on the futures curve, whether front-month or farther out.

3.7 The Rise of the Index

So far so good – but even while the method of maintaining a long-term investment in oil futures is now clear, as with physical trading this remains difficult to carry out in practice for most institutional investors. They generally lack the scale and expertise to manage such complex future positions themselves on an ongoing basis. This is where helpful investment banks come in. Starting back as early as 1991 with the launch of the Goldman Sachs Commodity Index (GCSI), banks and other institutions have created indices that track the performance of theoretically invested rolling positions in the futures markets of various commodities and at various maturities. Such an index is adjusted daily not just in line with the spot movement in commodity prices but also in line with the prevailing roll and collateral yields; in other words it is a so-called "total return" index.

With inception at a set value on a certain date, the growth or shrinkage of a particular index since then reflects the cumulative tripartite return that would have been realised to date on a set amount of capital invested in a real futures position at that original point in time. As to why index providers go to such trouble to maintain an index mimicking the performance of a theoretical, permanently rolled-over commodities futures position, the answer is simple. They wish to sell institutional investors derivatives based on the performance of this index, meeting the appetite of institutional investors who would like to maintain such a position but are not practically able to do so. And, by and large, they have been very successful. The GSCI is the granddaddy of all such indices, and set a pattern now widely followed in that it is actually a diversified commodity index – it aggregates a permanently-invested futures return across a range of commodities at the front end of the various curves, so investors have a one-stop solution for diversifying their commodity exposure.

Indices do exist for individual commodities, and indeed the GSCI and its successors such as the Dow Jones-AIG (DJ-AIG) diversified commodity index launched in 1998 are constructed on the basis of such underlying single-commodity indices. The GSCI, for example, has a roughly 36% weighting towards an index tracking front-month and nearby WTI oil contracts, a 15% weighting towards an index tracking Brent oil, a 6% weighting towards one tracking natural gas futures, 4% towards a copper index, and a 3% weighting towards each of wheat and live cattle futures. As well as forming the basis for the larger diversified index, these individual component indices will often themselves be used as the basis for products, both by the investment banks themselves and third parties. Many single commodity ETFs are created and managed, for instance, by companies such as ETF Securities who are independent of the actual index providers but nevertheless have an agreement to reference their products.

In terms of how the total return is calculated by the commodity index provider, it is generally presumed that the investment is passive, and long-only. In other words the notional position being referenced is not actively managed with regard to its return but is presumed instead to remain invested in that commodity for the duration regardless of performance, and also that these futures contracts are all presumed to be for buying rather than selling the commodity in question. So all else being equal, the index will increase as commodity prices rise and drop as commodity prices fall – as long as the roll return does not eat up this spot performance. It is also presumed, for the purposes of calculating the collateral return portion of the index performance, that the notional investment in futures is fully-collateralised; in other words not just a margin deposit but the whole value of the original investment has been put on deposit as collateral for the position and is therefore earning the 90-day T-bill rate.

Conditions like these, along with others vital to ensuring a consistent and predictable method of index calculation, are enshrined in agreements between index providers and customers accessing the index. This means potential institutional investors are now free themselves to decide if they want to bet on whether this replication of a passive, long-only front-month futures position will rise or fall; in other words whether they want to buy or sell the index in question. The easiest way for them to do so is to simply agree an OTC derivative swap of some sort, going either long or short of the index at whatever the prevailing level is for a set amount of capital at risk, with a broker such as an investment bank – in other words, a swap dealer.

Most institutional investors themselves, such as pension funds, have to date tended to go long of the indices, anticipating the positive return from tracking commodity price appreciation. The kind of commodity index tracker fund that is constructed in such a deal, invested in by financial institutions on the basis of a derivative tracking the performance of a specified index or combination thereof, is now seen as the largest single contributor towards commodity investment overall. As Hubard reports, Standard & Poor's (S&P) estimated 2007 total outstanding commodity index investment of at least $130 billion was up from $100 billion in 2006, and almost twice what it was in 2005, while at the same time total commodity investment managed by hedge funds was reckoned to be around $55 billion. S&P in fact bought the rights to the GCSI index from Goldman Sachs in 2007 because the latter investment bank and swap dealer felt that – given the amount of derivatives business it was writing on the basis of the GCSI family of indices – transparency required some distancing of itself from the provider responsible for calculating the index levels themselves.

3.8 Black Gold's "Dark Matter"

Overall, total outstanding OTC commodity derivative investment was reckoned at some $6.4 trillion by the Bank for International Settlements in mid-2006. This figure is for the notional capital at stake in all OTC derivatives, including those entered into by commercial commodity users for genuine business risk hedging, alongside purely speculative derivative exposure entered into by financial investors of the sort described above. Against this total, perhaps some $100-150 billion invested in commodity tracking funds does not seem like much at all. But alongside such tracker fund interest, speculative commodity exposure by financial institutions also involves instruments falling outside this description, such as single-commodity swaps (which might end up repackaged as exchange-traded funds) or structured commodity notes. All these products, too, are generally based on OTC swaps referencing the commodity prices or commodity index levels, much the same as the swaps underlying tracker funds.

A survey of financial sector investors in OTC commodity derivatives released in May 2008 by investment consultancy Greenwich Associates shows that these latter instruments are almost as significant in portfolios as index exposure. While 43% of total institutions polled had exposure to commodities via index swaps, 39% had exposure through single commodity swaps, and 21% through structured notes (these are overlapping, rather than mutually exclusive, exposures). It is clear that index funds are only part of the commodity investment story for institutional investors, and not even the most important part for certain classes of respondent. While 59% of investment and pension funds surveyed did indeed hold some index exposure, only 40% of banks surveyed did, but 80% of the latter had single commodity swap exposure.

In truth, then, no one really knows exactly how much of that $6 trillion-odd figure for OTC commodity derivatives is down to speculative financial investment, and the main reason for this is obvious. By definition OTC derivatives are unregulated, in general traded away from recognised exchanges (although some are in fact cleared through exchanges), and most of the trade in them is therefore not subject to the same sort of oversight as commodity dealing on recognised futures exchanges. OTC commodity derivatives are an informational "black hole". Yet if little light escapes these off-exchange markets, it could nevertheless be argued that the dealing that goes on in them is also the "dark matter" of oil price formation.

In the same sense that physicists postulate the existence of something they term "dark matter", a form of matter which they cannot actually see but which they feel must exist because only with this additional gravitational weight thrown into their sums do the orbits of galaxies match the physicists' preferred predictive models, it can be suggested that an explosion of unregulated, largely invisible OTC derivative trading must be what was moving the oil price when all of the visible factors are not seen as sufficient to justify where prices were in the first half of 2008. As Greenwich Associates themselves concluded in their May 2008 study, with regard to financial investment in OTC commodity derivatives specifically, 'While the long-term fundamentals of global energy and other commodities markets are being driven by increasing demand, there is little doubt that, in the immediate-term, speculative investors are driving up both trading volumes and prices.'

There is one objection raised time and again to this viewpoint. It is an objection that is so easily dismissed after a short recap of how OTC derivatives trading actually works as a business that it is surprising it is raised at all. But it is raised nevertheless and by even quite heavyweight commentators, so it is worth treating. It is, namely, that all these OTC derivative exposures to oil are just so many "paper barrels", even more so than the barrels traded on regulated futures exchanges – precisely because as these derivatives are purely financial swap agreements, they have no reference whatsoever to taking or making delivery of physical oil itself.

As such, this "only paper barrels" viewpoint argues that the weight and direction of this OTC derivative exposure on the part of financial investors can make no difference to actual oil prices because it plays no actual part in the futures market in which the actual oil price is discovered (supposedly on the basis of the fundamentals of supply and demand). At first glance, there seems some beguiling logic to this position, because regulated oil futures exchanges and unregulated off-exchange derivatives trading are indeed separate spheres of trading liquidity. So there is no obvious, formal connection between the two, such that volumes of OTC exposure off-exchange should count in the reckoning of prices quoted on-exchange, which reflect instead the volume of demand evidenced on just that exchange.

As we have seen, however, there is in fact a very real connection between notional volumes of exposure traded in OTC derivatives, and actual volumes of oil futures contracts traded on the regulated exchanges (and therefore contributing to price formation there). This linkage is of course the swap dealers themselves, who will as described above invariably go to the regulated, exchange-based futures market to hedge off the net exposure they end up holding as a result of unregulated, OTC derivative contract writing. Moreover, not only is it likely that the swap dealers will hedge this net exposure in the futures market, they are in fact expected to do so by none other than the CFTC itself.

We have already mentioned the fact that swap dealers are exempt from the position limits that restrain other market participants in futures exchanges, but this is not the full extent of the "swap dealer loophole" that is denounced by many experts as allowing financial sector speculation to drive oil prices. The loophole in its current form originates from 1991, when according to the CFTC's own account, a particular swap dealer enquired of it whether or not net swap exposures incurred specifically as a result of writing derivatives based on commodity indices could be cleared in the futures market in the same way, without position limits, as the exposures incurred in structuring derivatives for more obviously commercial customers. The CFTC ruled that they could.

3.9 Through a Glass, Darkly

Add up all the considerations above and it is clear that a mechanism does indeed exist for the translation of volumes of interest going long or short of oil prices at particular levels through off-exchange derivative agreements into volumes of interest going long or short of oil prices at comparable levels in the oil futures market. This is the needful and customary practice of the derivative-writing swap dealer hedging off the net exposure he holds (as a result of taking on the counterparty role in all these swaps) by taking out on the regulated future exchange the opposite trade to his own net exposure, and by implication mirroring the net position his own customers have taken in striking the swap agreements.

In other words, if in aggregate their customers have gone long of WTI oil, the swap dealer will have to go long of oil for the equivalent amount on Nymex; or if they have gone short, then the swap dealer will have to go short. And these volumes will in turn then be part of the whole set of pricing pressures leading to the oil price quoted in the daily papers. Moreover, there is of course good reason to suspect that in aggregate the tilt this speculative financial interest will add toward any swap dealer's net exposure will be towards the long side, adding to other upward pressure on prices from buying. This is because, as noted above, much OTC derivative investment by institutional investors is invariably going long of commodity prices, almost unquestioningly.

It is theoretically possible that if a swap dealer could find another dealer such as himself off-exchange, he could simply hedge away his own net exposure here by structuring a swap on his own account with the other dealer, without therefore having to go and deal in regulated futures. So he could; but that possibility depends on the likelihood of finding another swap dealer ready to take the opposite end of this trade, a swap dealer whose own net exposure is long commodities because his own customer book is in contrast full of parties going short of commodity prices. In reality this is unlikely to happen.

The Greenwich Associates' research quoted above cited two investment banks in particular as between them transacting the bulk of OTC commodity derivative writing, both for financial institutions and for the commercial players. These were Goldman Sachs and Morgan Stanley. When a market is dominated by just two players in such a fashion, they are not likely to have books showing significantly differing net customer exposures, with one being noticeably long of a particular market versus the other being noticeably short. The law of averages dictates that they will hold roughly the same exposures to roughly the same clients. So, facing only each other, and holding the same exposures, these big boys of OTC commodity derivative trading have nowhere else to go in hedging their exposure than the actual real futures exchange itself.

What difference does this new view of swap dealers make to the balance of speculative versus genuinely commercial interest in the Nymex crude market? It seems clear that, given the rise of institutional investor interest in commodity derivatives, this speculative element will have a significant role in reaching the net exposure a given swap dealer has to bring to Nymex to hedge away. But with the way data on this Nymex market segment has historically been collected and presented by the CFTC, there is no authorised estimate available of how much swap dealer activity on Nymex relates to hedging exposures due to financial speculation, as opposed to supposedly more traditional exposure to commercial hedge counterparties. We could, however, get some sense of what the outer limits of such a scenario would look like, by simply reversing the CFTC classification of swap dealers from commercial into non-commercial. And we can do this because, while the CFTC normally limits its reporting split in the standard CoT report to the aforementioned commercial/non-commercial, a couple of academic studies released under its auspices have recorded a more detailed breakdown of market participation.

Figure 8 recasts the same data for total Nymex WTI open interest data (shown previously) to demonstrate the split between commercial and non-commercial interest at three different points in time – 2000, 2004 and 2008. That data comes from a CFTC study released at the end of 2008 (by Büyüksahin, Haigh, Harris, Overdahl and Robe – see Sources & Bibliography), which also revealed the subdivisions within these two broad classifications at each of these three snapshots of market activity. In our figure we pull out various sub-divisions as identified in this data into their own category, so that we can both appreciate their relative importance to the overall growth of open interest on the futures exchange through this period 2000-2008, and also construct our own new index of speculative potential on the Nymex exchange. Under the new data split, we identify as specific trading groups the "swap dealers" themselves, the "hedge funds" (or MMTs), the other parties classed alongside MMTs in the CFTC non-commercial category as "other non-commercial", and lastly an "unquestionably commercial" category, which is simply the "commercial" category according to the CFTC distinction minus the swap dealers.

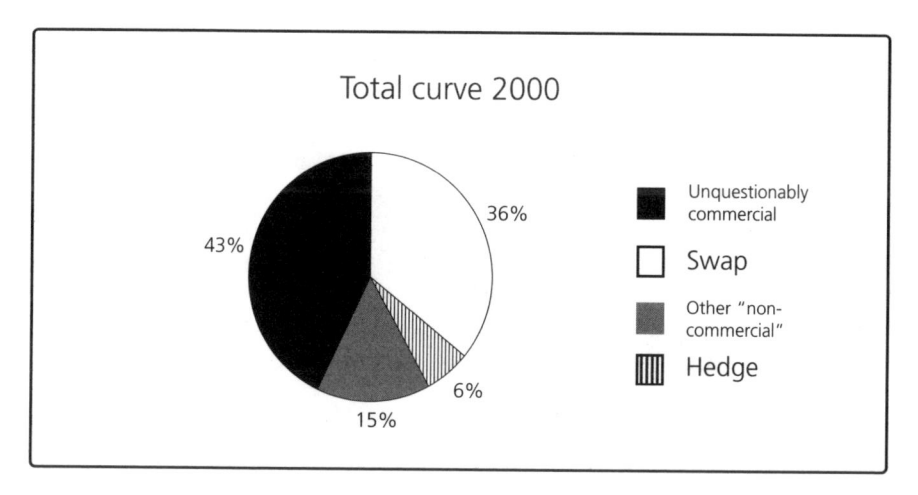

The Financialisation of Oil

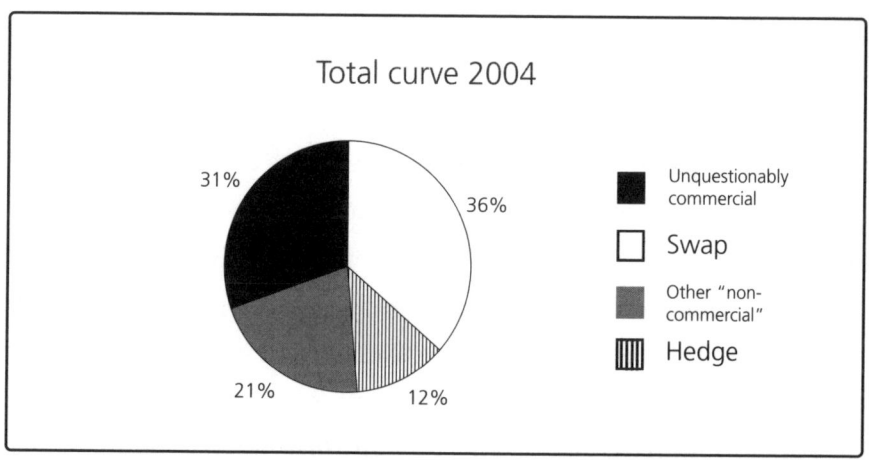

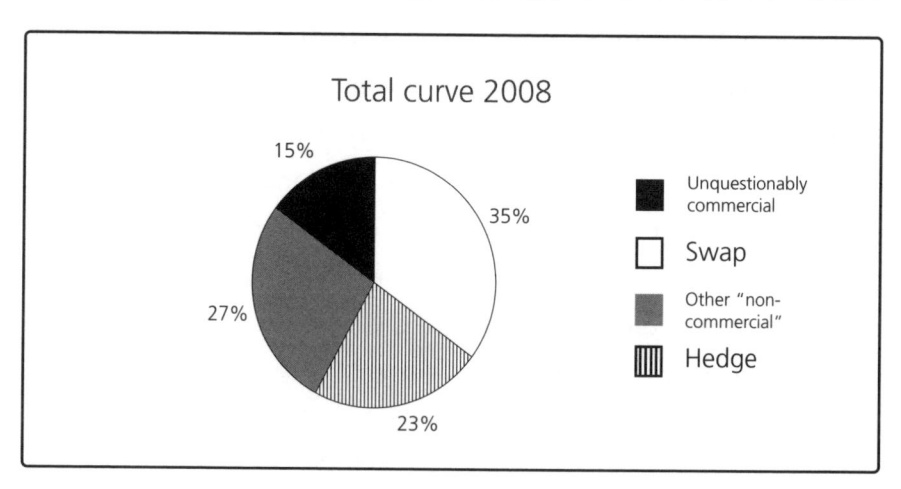

Total curve 2000

- Unquestionably commercial: 43%
- Swap: 36%
- Other "non-commercial": 15%
- Hedge: 6%

Total curve 2004

- Unquestionably commercial: 31%
- Swap: 36%
- Other "non-commercial": 21%
- Hedge: 12%

Total curve 2008

- Unquestionably commercial: 15%
- Swap: 35%
- Other "non-commercial": 27%
- Hedge: 23%

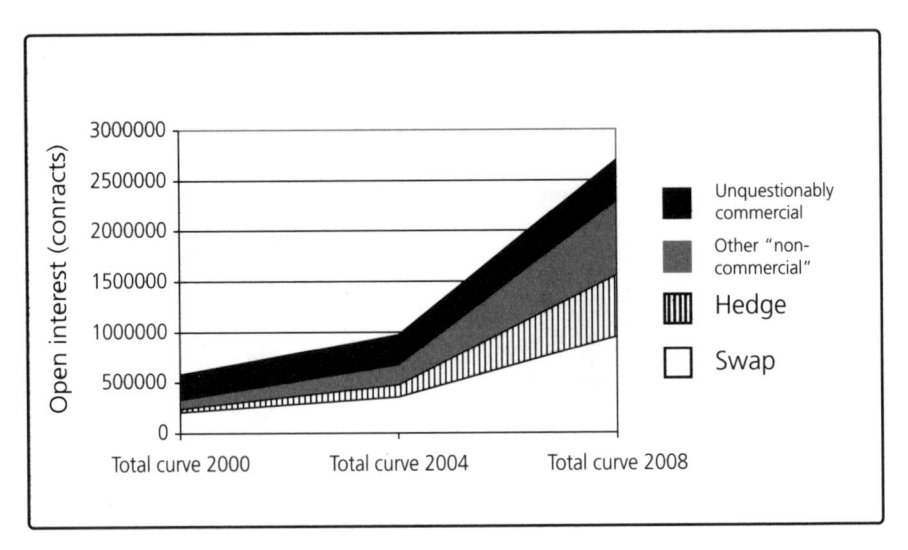

[Figure: stacked area chart showing Open interest (contracts) from 0 to 3,000,000 across Total curve 2000, 2004, and 2008, with categories: Unquestionably commercial, Other "non-commercial", Hedge, Swap]

Figure 8: Nymex oil market growth with more detailed trader definition
[Source: CFTC (Büyüksahin, Haigh, Harris, Overdahl & Robe, 2008)]

The first three categories added together give us our new definition of potential speculative activity on the market, as opposed to the unquestionably commercial activity. As can be seen, having reclassified the swap dealers into the potentially speculative category alongside the CFTC "non-commercial" traders, we can see that in all three years this new grouping dominates the market, but its dominance grows markedly over time. From accounting for 57% of open interest in 2000, this "potentially speculative" agglomeration accounts for 69% by 2004, and an incredible 85% by 2008. And of course, this really *is* incredible. With this example we are not seeking to say that by 2008 this actually *was* the proportion of open interest originating from financial speculation. We are, rather, saying that unfortunately this answer to that question is as valid as the answer given by looking at the data through the traditional lens of the CFTC CoT classifications. The truth undoubtedly lies between these two extremes – the problem is, no one knew where.

Nevertheless we can still draw other conclusions from this data. While the whole focus of the preceding passage has been on swap dealers, their actual share of the market is roughly constant at around 36% through these three historical samples. Of course, in absolute terms the volume this same percentage accounts for has grown massively, as we can see from the presentation of the same data in Figure 8. From under 250,000 contracts in 2000 (the actual exact figure is 208,638), it increased to around one million contracts in 2008 (the actual figure is 947,951). Nevertheless this realisation serves as an important corrective. The whole point about the swap dealer loophole is that it represents a significant additive to other speculative players already clearly acknowledged as such in the market, rather than replacing these players as the sole concern. Hedge funds – which in this data will mean those hedge funds acting directly for themselves in Nymex, rather than relying on floor brokers or swap dealers – display striking growth in market participation, from 6% to 12% to 23% for 2000, 2004, and 2008 respectively.

Meanwhile, floor brokers and traders – both the "market makers" and the "locals", for let us not forget that it was one of the latter who were supposed to have first pushed oil to $100 – also saw a huge increase in participation, from 15% to 21% to 27% through the three years. So rather than focusing on any one of these, we should instead observe that as these undoubtedly speculative players increased their market share, with swap dealers relatively static across all three periods, their growth came at the expense of our "unquestionably commercial" category. This latter category, the only slug of Nymex open interest we can (notionally) be sure is participating in the market on the basis of the physical fundamentals of supply and demand, saw its share of open interest shrink from 43% to 31% to 15% across 2000, 2004, and 2008. Whatever the exact proportion of all-important swap dealer participation reflecting exposure to speculative financial sector clients, the big fact to take away is that this period has certainly seen the balance of trade in the market shift toward speculative interest.

3.10 A Special Call

Among the expert witnesses at the various US hearings on the effect financial speculation had in driving price movements in the oil price, several lined up to concur with this sort of argument. It was argued vociferously both that more attention should be paid to swap dealer hedging in futures markets as a function of financial investor speculation, and that the "swap dealer loophole" that theoretically allowed this off-exchange financial investor interest to contribute to, and therefore distort, futures price discovery on-exchange, should be revoked and closed. Despite the paucity of data on the issue of swap dealer exposures, one witness at least attempted to put some figures to the legislature and the public at large. This was hedge fund manager Michael Masters, who we have already mentioned.

It would be fair to say that Masters became something of both folk hero and folk devil as a result of his prominence through a series of Congressional hearings. Eventually commissioned by US senators to prepare a report on swap dealers and their influence on the market, Masters released this report in autumn 2008 under the title *The Accidental Hunt Brothers*, and set up a blog of the same name to further promote his own take on the oil price blow-out. Yet this has not been without challenge – a prominent oil sector analyst, Philip Verleger, has dismissed Masters' work as 'junk analysis' (of which more later). For now, however, we are merely interested in one aspect of Masters' various Congressional presentations – his attempt to calculate how much Nymex open interest volume the commodity index speculators might account for when swap dealers come around to calculating their net exposures for hedging on that market.

In testimony submitted on May 20 2008 to the US Senate Homeland Security and Governmental Affairs Committee, Masters reckoned that index speculators alone would account for 31% of long-only interest in outstanding WTI oil contracts on the Nymex market. That figure can be compared with the total 36% of interest overall we ascribe to swap dealers in 2008. Was most of this then obviously speculative financial interest on the long side only? Not necessarily. Obviously, if the swap dealer had managed to pair off such a large exposure to long-only commodity index investors, with an equivalent weight of other OTC derivative interest going short of the oil price, then its net exposure to hedge would be zero. Or if it had twice as much short interest as long then this would spell a net short exposure of equal magnitude. Yet the absolute vacuum of data we have already commented on with regard to total OTC market positions means there is nothing to gainsay the implication that just index investment alone in OTC derivatives – i.e. not including single-commodity swaps or exchange-traded funds or structured notes – could be a serious weight in swap dealer exposure to clients going long of the oil price, and therefore a serious weight in swap dealer interest in buying oil futures.

How had Masters reached his figure for the proportion of long open interest on Nymex that should be attributed to speculative index investment? Ironically, while the crucial data regarding the size of Nymex crude oil open interest attributable to index speculators was at that time unavailable, the CFTC did actually maintain such data on the volume of speculative index trade swap dealers brought to hedge on Nymex when it came to agricultural commodities. Perhaps more sensitive to the charge of allowing undue speculative influence over the historical core of the futures market, here the CFTC had compiled for some time a record of the volume of open interest in these contracts due to index speculation.

Masters extrapolated from these CFTC disclosures the dollar value of these contract positions, and then satisfied himself that these figures tallied roughly with the stated weight of these commodities in key commodity indices such as the SP-GSCI and the DJ-AIG versus the total dollar value as stated for these indices by their own providers. The method seemed to work, so Masters then worked backward from the same total dollar value of these indices to find the contract-equivalent open interest weight he should ascribe to index speculation in the thirteen commodities still not covered by the data on agricultural elements. Notably, while Standard & Poor's estimated total commodity index investment sat at $130 billion, Masters himself reckoned that total commodity index investment in March 2008 had grown to $260 billion.

The CFTC had its own response to Masters' evidence and the picture it painted. Nine days after his testimony above, on May 29 2008, the US futures market regulator announced it was making a 'special call', an enquiry into market participant behaviour with compliance backed by legal sanctions. Swap dealers were to open up their black boxes of net exposure calculation and reveal just how much speculative financial investment was contributing to their overall net positions as hedgers on Nymex. The scope of the information sought was as follows:

- How much total commodity index trading is occurring in both the OTC and on-exchange markets?

- How much commodity index trading is occurring by specific commodity in both the OTC and on-exchange markets?

- What are the major types of index investors?

- What types of clients utilise swap dealers to trade OTC commodity transactions?

- To what extent would the swap clients have exceeded position limits or accountability levels had their OTC swap positions been taken on exchange?

In a letter dated June 11 2008 to Senator Jeff Bingaman, the chair of the US Senate energy and natural resources committee, CFTC Commissioner Walt Lukken explained the brand new thinking in the agency regarding swap dealers:

> The Commission is using its existing Special Call authorities to immediately begin to require traders in the energy markets to provide the agency with monthly reports of their index trading to help the CFTC further identify the amount and impact of this type of trading in the markets. In addition, the Commission will develop a proposal to routinely require more detailed information from index traders and swaps dealers in the futures markets, and review whether classification of these types of traders can be improved for regulatory and reporting purposes. Lastly, the Commission will review the trading practices for index traders in the futures markets to ensure that this type of trading activity is not adversely impacting the price discovery process, and to determine whether different practices should be employed.

In his letter Lukken also referred to the CFTC announcement one day previously, on June 10, of the formation of a new US government 'inter-agency task force' to examine the whole question of speculation in commodity futures markets. This task force would include staff representatives from the CFTC, Federal Reserve, Department of the Treasury, Securities and Exchange Commission, Department of Energy, and Department of Agriculture, and would examine investor practices, fundamental supply and demand factors, and the role of speculators and index traders in the commodity markets. As Lukken wrote, 'It is intended to bring together the best and brightest minds in government to aid public and regulatory understanding of the forces that are affecting the functioning of these markets, and it will strive to complete its work quickly and make public its results.' The CFTC was finally going to address the central questions raised in the hearings – light was going to be shone on the dark matter of oil futures trading.

4

The Peak Weeks

'Commentators seeking an explanation for the unprecedented surge in crude oil futures on Friday – up by nearly $11 a barrel – have an embarrassment of riches. The spike (or is it a new plateau?) can be blamed on everything from short-sellers covering themselves, to hawkish comments from an Israeli minister, to a slump in the dollar, or just the catch-all epithet "speculation". The only kind of speculator who would seriously push up the price of oil today is the kind who holds physical stocks of oil away from the market, hoping to sell later when prices are yet higher. If such speculators exist, either they are Opec governments, or they are doing a splendid job of hiding their inventories, because the data suggest that end-users are burning all the oil they buy. It is possible that speculators are indeed driving up the price of futures contracts, but it is hard to find much evidence that they are bidding up the price of the black, sticky energy source that goes into oil refineries. Efforts to fight speculation are at best a distraction from the real problem: oil supplies are not growing much, while oil demand still is.'

Editorial leader, *Financial Times*, June 9 2008

'Outside of the US, we have little doubt that there is an ongoing build in inventories this quarter. But it appears to be taking place far more in China and the Persian Gulf than in the OECD and, as a result, is less visible and difficult to measure.'

Edward Morse et al, Lehman Brothers, "Oil dotcom", May 29 2008

4.1 Entrenched Ideas

Despite the logic of the arguments regarding swap dealer hedging requirements in the futures market, through May, June and into July 2008 it was clear that the idea that speculative factors could be occluding real world indicators in driving breakneck oil price appreciation was widely derided. Such derision came not just from the bodies of regulatory officialdom such as the CFTC, but also from numerous influential organs of opinion. As just some examples from many that could be quoted, mere days before crude clocked its all-time peak, Britain's two most august and best-known business newspapers ran back-to-back dismissals of arguments attributing sky-high oil prices to speculative pressures. Following on from the editorial column quoted above, July 4 saw the high-visibility "LEX" column in the *Financial Times* declare of complaints against speculators, 'This simplistic argument blithely ignores fundamental supply and demand issues.' And in its July 5 issue *The Economist* noted, under the headline 'The oil price: Don't blame the speculators', that 'This reasoning holds obvious appeal for those looking for a scapegoat. But there is little evidence to support it.'

In maintaining this stance the newspapers were lining up with politicians on both sides of the Atlantic. Part of then-US Energy Secretary Samuel Bodman's case for haranguing the Saudis to increase crude production in May and June was that, as he repeated to reporters on the eve of an energy conference in the Saudi port city of Jeddah in late June, 'There is no evidence that we can find that speculators are driving futures prices.' Also in June, the UK Treasury published a report which noted of financial speculation, 'Although there is insufficient evidence to conclusively rule out any impact, it is likely to be only small and transitory relative to fundamental trends in demand and supply for the physical commodities'.

Hindsight is a wonderful thing, but why so many commentators were to be proved wrong on the supposed durability of triple-digit oil prices is itself a key part of this story that bears examination. After all, one thing guaranteed to further inflate a speculative price bubble is potential but as-yet uncommitted investors being told repeatedly and with the reassuring gravitas of accepted authority that there is in fact no bubble after all. What were the arguments deployed to bolster an "establishment herd" of regulators, analysts, journalists, and politicians in a shared view that speculation was not a key factor to consider in analysing oil price movement? Some were without any merit whatsoever; others were more cogent. We shall examine a few in turn, from the fatuous to the more convincing.

"For every buyer there is a seller."

It is surprising that we have to consider this as an argument at all, but nevertheless it was repeated so many times throughout this period that we must. We have already described the pricing pressure that the new wave of speculation would bring to bear on futures, emerging through the financialised oil market described above; and, further, how this would, in light of the ruling grand narratives and the long-only bias these investment stories encouraged in financial investors, invariably fall on the long, buy-side. The "for every buyer there is a seller" argument saw this as irrelevant because each speculative buyer is matched by a seller, so there is as a result supposedly equal pricing pressure in the opposite direction from those selling. The implication is that speculators cannot move the market full stop. But this is obviously balderdash. If it were not, how would market prices ever move in any direction, whether up or down?

The point isn't whether or not a buyer finds a seller. For a transaction to have occurred, and register as the last price for a particular contract on Nymex, by definition they must have done. The important issue is at *which price* the seller has had to go short against the buyer, which is in turn linked to the volume of buying interest present in the market. For argument's sake let us presume that an investor comes to Nymex seeking to buy 150 contracts for delivery of crude in June 2012. On the offer side of the market, the current best quote is $133, but for 50 contracts. The next best quote is $133.30, for another 50. Next in the stack of offers are a lot of 100 contracts, priced at $133.50. As the single buyer snaps up all the contracts he requires for his order, he will completely eat through the first two offers, and halfway into the third. All else being equal, by the time he has finished, the best offer in the market will now be 50 contracts, priced at $133.50. So on the offer side of the equation, the price will have moved from $133 to $133.50, and the volume of contracts available at this price will have dropped to half what it was prior to this transaction.

Seeing what has happened, other market participants will adjust their quotes accordingly. While many would move theirs further upwards, some may well come in with a new offer below $133.50. But as vendors, they would have to have a good reason for offering at a price lower than the previous transaction, and normally this would only happen if there was no apparent buying interest, which in our example is clearly not the case. Volumes of interest on either the buy or sell side in a quoted market move both actual prices for the last transaction and also the quoted prices around the bid and offer side simply by taking out volumes previously available at certain prices on the bid or offer side. This is not rocket science but a basic principle of how all paper markets for quoted assets function. So it is amazing that people who happily accept without question this logic in markets such as those for listed company shares seem to have a blind spot for recognising it at work in another paper market: oil futures.

"Paper barrels cannot move the oil price."

Not so obviously fatuous as the first argument above, but given the preceding chapters the reader will hopefully now understand precisely how paper barrels do indeed have the power to move the oil price, in the sense of both the prices of oil futures themselves feeding into the physical spot price, and the wider world of OTC derivatives feeding into the prices of oil futures through swap dealer hedging requirements.

"Nymex is too liquid and efficient a market for speculators to dominate."

This is an argument heard many times from various analysts when discussing the effect of speculation on the oil price. It is simply an affirmation of the CFTC line that while there is obviously a speculative element at work in the oil futures market, it is subordinate to and indeed, as averred in the CFTC study on hedge funds, follows on the coat-tails of a market moved by genuinely commercial parties on the basis of the fundamentals of supply and demand. Such an opinion is very much justified in relation to your own take on the volume of speculative interest in the market, the effect this has on prices, and whether or not the oil price as a result ever loses touch with the fundamental realities of supply and demand. This book as a whole addresses all these issues and the argument above will stand or fall with your opinion after reading it.

We should note, however, that belief in the efficiency of markets such as Nymex in achieving the optimal possible allocation of capital to resources is, for many commentators, more than a contingent judgment passed on this particular market or that. It is, rather, a cornerstone of the dominant type of contemporary orthodox economic thinking, which in its underlying attachment to the market mechanism as the ultimate arbiter of value goes so far as to deny the very possibility of such a thing as a speculative bubble affecting the prices discovered through its workings. We shall look again at this conditioning belief in a later examination of the pathology of speculative bubbles; however, for now let us say that for most modern financial analysts trained in this thinking, saying that speculators cannot determine price movement in a market is as much a religious assertion as a judgment based on observation.

Like much religious thought, such belief in perfection can lead to amusing double-think. Anecdotally, I can recall having the discussion above with an analyst in late May 2008, establishing that his opinion was indeed that it was impossible for speculators to outweigh fundamentals on the Nymex. Fair enough. I then enquired as to what the analyst thought was behind a particularly striking shift in the futures curve – from backwardation into a perfect contango – which had just occurred. This, said the analyst, was due to a large volume of non-commercial market players taking off a time-spread trade that had gone wrong. The speculators had all gone long of the front-month and short the back end of the curve. The trade had moved against them and many had moved to unwind the position.

As such they were selling the front-month and buying the back end; the front-month contract dropped, and meanwhile the back end contract in question was so comparatively illiquid that the large volume of offsetting buys coming into the market caused its price to shoot up. Hence the comparatively rare instance of the curve jumping into a full contango, with prices rising steadily out to the farthest contract. I have also heard two other versions of what happened that same week in May 2008, which we shall also examine in due course. For now, however, readers may have noticed: on the one hand this analyst was saying speculators could not move the market, and in the next breath he was describing in detail how a particular group of obviously speculative market participants had not just moved a price but the shape of the whole curve.

"Commodities not traded in futures markets have appreciated just as much."

This argument asserted that the rise in futures-traded commodities such as oil was not due to speculative pressure, because its appreciation matched that seen in other "bulk" commodities like iron ore, ferro metals and coal. For whatever historical reasons, these commodities are not continually traded on computer screens in futures markets to determine daily prices, but are instead characterised by bilaterally-contracted prices holding through set periods. As such they are seen as largely impervious to speculative financial investment. But prices for these commodities had indeed shot up in tandem with futures-traded commodity market bellwethers such as oil or copper, and the reasoning above held that because the increase in oil prices did not, by some calculations, look out of line with these increases, the oil price appreciation was as fundamentally-based as that seen in iron ore, which indeed had a spectacular ascent of its own through recent years.

This argument sounds reasonable enough, but fails ultimately for an obvious reason. In truth, it merely shows that, if speculators have moved oil prices, in the absence of finding support in physical fundamentals they have simply looked to observable price appreciation in other, bulk commodities, as a rough yardstick for a price level beyond which their own buying spree might start to look "toppy", as the lingo goes. In reality, the physical supply and demand fundamentals across the wide range of bulk and screen-traded commodities vary so much that it is impossible to reckon some sort of universal level of price appreciation that looks "appropriate" across all of them.

4.2 The Serious Competition

Beyond the relatively lightweight arguments against speculation playing a major part in price determination outlined above, the establishment view also had in its favour three more heavyweight lines of reasoning which all deserve serious attention. We can describe them as follows.

"Speculation in oil futures is always accompanied by hoarding."

Some clichés are so because they are true, and the sentiment above is grounded in the reality of oil trading. When prices further out along the futures strip offer a return in excess of the simple cost of storage through the same period, it encourages people who have access to physical oil to place it in storage while locking in the gain by selling the same oil forward through contracts on the futures market. This is in fact the practice that most people would most readily associate with "speculation" on the oil market, and its obvious corollary is that oil in storage tends to rise when participants are engaged in this play, or there is "hoarding" as the common term goes. Therefore, goes the reasoning, no visible hoarding means no speculation is at work.

We have already noted how, in fact, through this period oil inventories were actually building. At the time, however, this was not so obvious to many market participants – oil stock data is invariably backward-looking, and sometimes bears considerable revision following its initial appearance. So let us allow that many certainly could not see any visible hoarding through this period. Nevertheless, this argument still would not hold any water against speculation in the sense we have specifically described throughout preceding chapters. Note that this sort of speculation accompanied by hoarding is only possible for those with access to physical oil – but that the very definition of speculator we use for futures markets is, *pace* Nymex, someone without a commercial, physical interest in the commodity at hand.

Long, buy-side demand from speculators in oil futures would not result in any hoarding on their part, at least. What about those selling it to them on the exchange? Surely they need to hold stocks against the risk of having to meet a physical oil delivery if they cannot offset their way out of this commitment at a reasonable price? Would not hoarding on the part of contract sellers to meet physical delivery on a futures contract sold to a speculator still betray the latter at work? Not necessarily. Firstly, it is presuming that the sell-side on a futures contract will always be a physical market player, and while this is historically often the case it is not a foregone conclusion. There is nothing to stop a non-physical, non-commercial, financial speculator in the market from selling as well as buying futures contracts, in which case there will be no visible hoarding attached to their position in this transaction.

Moreover, even if the seller of the contracts the financial speculators are buying is indeed a physical market player, they may yet forego the security of having oil in storage to back their commitments if they feel that the likelihood is that these commitments will not be called on any time in the near future. This might be the case if they understood that a large part of volume on the buy-side in the market is committed to rolling out of taking actual delivery every month and instead being permanently invested in the front-month – which means that those selling into this demand will always find a liquid market in which they will be able to buy back their own equivalent contract commitments prior to monthly delivery as well. As we have seen, this is exactly the situation that obtains with index investors. Further, as this index money is long-only, it will remain automatically committed to rolling over every month no matter how negative the resulting roll yield may be – so even when the market is in contango and the index investor is suffering, this trade remains on the table to be rolled over every month.

So much for the front end of the curve, but as we have seen financial speculation in oil futures is not limited to index investors piling into the front-month. There are institutions investing in long-term single commodity swaps; institutions buying medium-to-long-term structured commodity notes; institutions buying exchange-traded oil funds which are not limited to front-month positions, but also allow investors to be permanently long of the two- or three-year price; and institutions buying dated options for various calendar points far out along the curve. All these forms of investment would see swap dealers hedging out client exposure by buying Nymex futures at the corresponding medium and long-term dates along the curve. Even if the originating counterparty to these trades is a physical player, they are unlikely to right now put in storage barrels that will not be called on for five years – when in the interim the physical player, particularly if they are a producer, will have run through several more inventory cycles themselves.

They are much more likely to rely on their presence in the physical trade to secure the requisite barrels far closer to the contract expiry date than to tie up capital and pay inventory storage costs from now. The idea that speculation is always accompanied by hoarding is fine if describing a typical tactic of physical market players holding barrels to realise locked-in profits within months, rather than years. (And we have in fact seen episodes of such hoarding accompanying this kind of "old-fashioned" speculation by physical traders a couple of times since the oil price crash of late 2008.) It fails, however, as a cast-iron rule when facing the realities of the new forms of speculation now at work in the futures market.

Diesel fundamentalism

Asian hunger for diesel had to be considered in any reckoning of demand-side pressure in the oil markets. As previously explained, the twist diesel brings to the story is that all oils are not created equal. Light, sweet crude is favoured over heavy, sour crude. The implication is that the market overall might have been balanced in appearance, but that in reality the crucial light sweet crude was undersupplied versus demand. As WTI oil itself is the premium light sweet blend, its price – both spot at Cushing and by extension front-month at Nymex – was directly subject to a demand squeeze not noted for other blends of crude, particularly comparatively sour output from several large OPEC producers. Thus while OPEC might be finding it hard to sell its crude – and toward the peak of the oil price, a veritable flotilla of Iranian tankers full of sour crude were indeed conspicuously floating out at sea, with no demand for their cargo – the high "oil price" as quoted on Nymex did indeed reflect high demand, but for exactly what it referred to rather than all crude oil in general. And so it was genuine real world fundamentals driving the oil price after all, rather than speculation.

This line of reasoning has to be taken incredibly seriously, not least because one of its proponents, oil sector analyst Philip K. Verleger, had testified before Congress in December 2007 that oil would shoot up as far as $120 per barrel in 2008 as a result of this light sweet squeeze. Verleger's argument was, however, that this squeeze was being made worse than it had to be for quite artificial reasons – US government insistence on filling its Strategic Petroleum Reserve (SPR) with too much light sweet crude oil. The US Department of Energy (DoE) had decided to add to its strategic reserves and had commenced stock-building in August 2007, continuing through the rest of the year and into 2008. The SPR was being filled with a mix of sweet and sour crudes, but Verleger reckoned that even though the quantity of light sweet crude the government was taking out of the market amounted to less than 1% of the total global light sweet crude supply (of perhaps 10 million barrels per day), it was nevertheless sufficient to push prices upwards given the tightness in the light sweet market.

Verleger had reached this conclusion by rejecting as causes for oil price appreciation all of the other "fundamental" reasons we have enumerated previously, but also by rejecting, too, the idea that speculation was the cause. Testifying in December 2007, he calculated that commodity index investment in crude oil had actually declined in the preceding couple of months, so it could not be behind the price rises from around $70 in late summer 2007 to near $100 by the end of that year. If it was not the "traditional" fundamental factors, nor the speculators behind the price rise, it had to be the only other factor that had changed in 2007 – the DoE policy of removing light sweet crude stocks from the market through the latter half of the year. By January 2008, US politicians were echoing Verleger's analysis and calling on the DoE to suspend light sweet crude transfers into the SPR. The government initially refused to budge and it was not until May, when senators had finally passed a veto-proof law to end the transfers, that the DoE capitulated and said its current programme of stock-building would end with the transfers already scheduled for June 2008.

Verleger's arguments invited some criticism because he asked a lot of heavy lifting in terms of price appreciation from a very negligible difference in market supply. Yet his conviction was that if there really is no more light sweet crude available at the margin of supply, even a small supply disruption can cause a significant price spike. He estimated this effect at between an extra 25 to 40% price increase per 1% supply shortfall in light sweet crude oil. Verleger did, however, also muddy the waters somewhat himself. He felt that hedging by options derivative writers on the Nymex market could also be putting upward pressure on prices – but he saw this hedging as originating against exposure to commercial market players, specifically airlines hedging their oil price exposure, rather than speculative investors.

But it is worth noting that speculative investors would have exactly the same effect if it was they who were buying long options on the oil price from the derivatives writers instead of or even alongside commercial players. The derivatives writers themselves would still have to go long of oil by buying equivalent options on the Nymex. In media comment through 2008, Verleger himself also made repeated references to the weight of speculative investment in the oil market as contributing to high prices. Yet after the oil price had collapsed by autumn 2008, he was scathing, in particular, of Michael Masters' arguments, and dismissed the idea that speculation had played any part in the price blow-out. He seemed to be back to a light sweet crude fundamentalist view of the whole episode.

Whether or not DoE stock-building was artificially tightening the light sweet crude segment of the oil market to a noticeable extent through the first half of 2008, there was certainly evidence by early summer that the market thought diesel itself was in short supply. The "crack spread", the margin refiners gain on splitting various products out of barrels of crude oil, on diesel-type fuels – or "middle distillates", as refiners know them – had widened considerably. This was a signal that if prices for oil in general had shot up, prices for diesel refined out of that oil had shot up even more. Asian demand-driven super-cycle theories had a ready answer – this was fundamentals *par excellence* at work. The question was whether the bulls were right and this new pressure on light sweet crude was now a permanent structural consideration in the oil price, or was this diesel demand predicated on something more ephemeral that might yet evaporate in short order? In due course we shall examine arguments that the latter might indeed have been the case.

"Future fundamentals require demand destruction."

It is appropriate at this point to refer to the theory that lay behind the high-profile calls of the Goldman Sachs analysts that became so identified with the oil bull case through the first half of 2008. It will already be apparent that in coming up with any case justifying the rapid appreciation of oil prices in 2008, some recognition must be made of the headline fact that many signals from the physical market simply did not support the idea that demand was currently undersupplied. We have seen it done above in the diesel fundamentalist argument (which argues that these physical signals fail to reflect real tightness in a particular segment of demand). Goldman Sachs indeed acknowledged this diesel case, and made it part of its analysis. But it preferred to foreground another element – the idea that high prices might not be responding to an undersupplied market today, but instead reflect legitimate, fundamental concerns about supply in the relatively near future. The transmission mechanism for these concerns to the oil price quoted daily in the newspapers was high prices at the back end of the futures curve dragging up the front-month price after them.

Jeff Currie and his commodity analyst team at Goldman Sachs set out its case in detail in its "Energy Watch" research note of May 16 2008, subtitled "A lesson from long-dated oil". The note is upfront in admitting that it has to explain 'a market that seemingly defies fundamentals', and then does so by refocusing attention from the front-month contract that feeds directly into spot prices, to the five-year forward price as traded in Nymex contracts. This in itself is relatively uncontroversial. Many oil analysts would agree that the five-year end of the futures curve, sufficiently removed from volatility at the front end of the curve caused by relatively short-lived and therefore non-fundamental supply shortages or gluts, may be the better guide to reckoning what the market thinks the "fair" price actually is for oil.

In its analysis, Goldman takes the five-year WTI Nymex price as revealing the true "structural" price for oil, and says that this has been rising as a result of a "structural re-pricing" ongoing in the market. According to Goldman, legitimate concerns about the adequacy of potential supply development to meet the current trend of demand growth were causing market participants to bid up five-year prices towards the level where they would start to force a reduction in demand growth, such that it moves back in line with supply growth capability. Only when the market had reached this equilibrium would prices stop appreciating.

Goldman reckoned that to force the trend in oil demand growth of 3.8% annually – this figure being seen simply as equivalent to then-forecast GDP growth – back in line with global oil supply growth trending at just 1% annually, a 14% increase in the long-dated oil price from its current levels was required. This meant that Goldman Sachs forecast a $149 per barrel price for five-year Nymex WTI by the start of 2009. As Goldman claimed had been the case with another "structural re-pricing" seen in 2004-5 (at the end of which five-year oil had settled at around $70 per barrel), they now argued that this rise at the back end of the curve had fed through into price rises at the front end of the curve. 'After remaining stable for more than two years,' the analysts wrote, 'the long-dated oil price (five-year forward) is once again driving the oil market. And like 2004-5... the rise in long-dated prices drags spot prices to ever-higher levels'.

With this dynamic at play, Goldman also reckoned that despite widespread dismay at the volatility in front-month prices from late 2007 onwards, these prices had in fact displayed normal cyclical, seasonal patterns as long as it was understood that they had done so not in relation to a flat yardstick but to a five-year price that had been constantly appreciating while these cyclical peaks and troughs played out. As Goldman thought it had the handle on the "time-spread" logic under which these cyclical front-end prices moved, versus the long "structural" price, it also hazarded a forecast for the front-month WTI price – through the second half of 2008 it would average $141 per barrel.

Goldman claimed that in 2004-5, the new equilibrium for long-dated oil had settled at a level that incentivised new, costlier production from unconventional resources such as oil sands to meet prevailing demand growth. But the analysts foresaw no new technological breakthrough this time, and the fundamentals of supply and demand meant supply growth could not be accelerated. This was due to what Goldman called 'the revenge of the old political economy', which it labelled as 'resource protectionism'. In other words, precisely the sort of mercantilism and resource nationalism we have already described as one of the grand narratives dominating the physical oil market (academically, mercantilism is recognised as the oldest distinct variant of "political economy" considered as a system of statecraft).

In this respect, Goldman ended up pretty much in the "pragmatic peak oil" camp – supply would not go much higher, not because of geological but political restrictions on extraction of fresh resources. So rather than incentivising new production that in reality could not be accessed, this time around prices would have to rise to a level that actually destroyed demand sufficiently for supply to balance it.

4.3 The Reality and Implications of Curve Integration

In its focus on the physical and political restrictions on supply growth versus galloping demand as the dynamic forcing oil prices ever higher, and in ascribing actual numbers to both of these and demonstrating the mathematical formulae for calculating the price rise required to crimp demand growth sufficiently, the Goldman Sachs analysis was resolutely based on supposed fundamentals. It also, however, touched on an intriguing issue – was it true that long-dated futures prices could drag front-month prices up and down? It is important to bear in mind how the futures curve is constructed: with the last traded price for each contract in question, even if this contract has not traded on the day in question. There is no necessary linkage between front-month and other prices. It is possible for the front-month to move a lot on a single day while another contract will not have traded at all.

In practice, however, what tends to happen is that trades on one part of the curve do have an effect that ripples through other contracts. If market participants decide there is sufficient read-across in terms of supply or demand signals to the price of oil in contracts they are invested in elsewhere along the curve, they may adjust positions there up or down as well. This might seem obvious enough but, in fact, for most of the period since oil futures started trading in earnest in the early 1980s, there was an observable disconnect between the front-month and the rest of the Nymex curve. And for most of this period, the rest of the curve rarely meant more than the one- and two-year-out prices, it taking time for the market to develop to the level it is at today where longer maturities are readily offered.

A CFTC study released in February 2007 noted that in the early 1990s, the vast majority of oil futures traded had a maturity measured in months, and growth in contracts longer than three years did not begin until 2004. The same study also demonstrated that while until 2000 there was statistically no reliable relationship holding between movement in the front-month and one- and two-year prices, in 2002 and 2003 some measure of correlation emerged, and in mid-2004, these three maturities achieved meaningful 'cointegration'. This means they started to move in a statistically-detectable "lockstep", maintaining a quasi-stationary position between themselves while moving up and down together. The CFTC's own research on the matter therefore suggests that front-month and longer-dated contract prices do exhibit meaningful linkage across their movements.

Importantly, the statistical proof of such cointegration makes no meaningful distinction as to which price might be the "first mover" and which the "followers" – indeed, within the terms of reference of such a study, this question does not make sense. We cannot say if long-dated prices follow shorter-dated, or if shorter-dated follow longer-dated, or indeed if they may switch around, one leading and now following. All we can say is that, for significant stretches of time, they seem to move together. The implications of this idea are considerable. On the one hand, if it is true that long-dated prices can influence the front-month price, then a major problem for those who saw high oil prices as driven by fundamentals was solved. It did not matter if actual day-to-day market conditions were not displaying any actual tightness that might justify soaring prices. As long as there were strong fundamental grounds for bidding up oil prices some distance out along the futures curve to a certain level, the movement across the whole curve is being triggered by fundamentals.

On the other hand, if one presumes that long-dated price movement can drag front-month prices around with it, it also makes the speculative case for oil price appreciation considerably deeper. Many of those who denied the speculative case did so because they did not, in their opinion, see enough speculative weight in the market in the front-month and nearby contracts alone to be materially capable of shifting prices at this end of the curve. In reaching this conclusion they typically focused on just the commodity index investors, the obvious class of financial speculator investing in the front end of the curve. But if we allow speculative pressures further out along the curve to also bear some weight in pulling front-month prices up or down, then the scope of interest that has to be considered in calibrating potential effect is far wider. It would take in all those institutions placing bets further out along the curve through the plethora of swap-based instruments we have already enumerated.

Tellingly, while the February 2007 CFTC study could not determine which price was leading which, it did attempt to determine which class of trader it was whose trading across these different maturities was most responsible for bringing their movement into cointegration. Its clear answer was – the swap dealers. The very market agent most associated with the transmission of speculative financial interest into the market is also the market agent the CFTC's own research suggests is responsible for clear linkage emerging between price movement across different maturities. So in this sense, Goldman Sachs might be right, but for the wrong reasons. In other words, it may be correct in seeing the front end of the curve as being dragged around by movement at the back end, but incorrect in attributing this movement at the back end of the curve to fundamentals. The whole phenomenon, encompassing both back-end prices pushing ever higher and also the front-end responding to this movement, might instead be down to speculative interest, transmitted primarily through swap dealers.

This is particularly the case as the farther reaches of the futures curve, particularly beyond the three-year window that most commercial users and consumers of oil limit their hedging strategies to, are particularly illiquid compared to the front end of the curve. They are therefore prone to move more with comparatively smaller volumes of trade, while also being favoured more by speculators than physical market players. The same 2007 CFTC study also does indeed acknowledge that at that point in time, significant commercial market players such as refineries had practically no exposure beyond three years, while swap dealers were a driving force in such long maturities. And this picture tallies with subsequent data, particularly the CFTC study released in late 2008 which we have already used to generate our alternative view of the split between unquestionably commercial and potentially speculative market participants.

As Figure 9 shows, on the basis of this data, by 2008 swap dealers held near enough 60% of contracts with maturities greater than three years, with hedge funds the second-largest group of traders holding such long-dated maturities, accounting for some 24%. Looking at the three snapshots of trader segment interest across these dates, another wider conclusion germane to our whole enquiry is unavoidable.

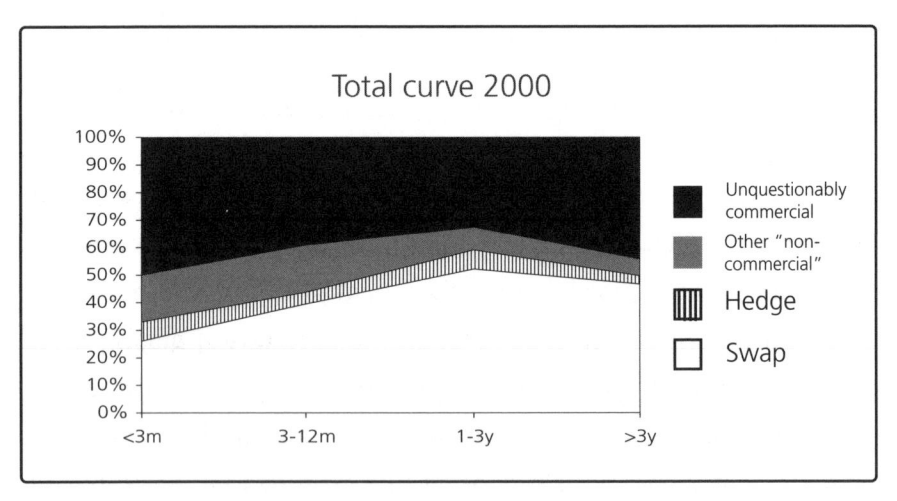

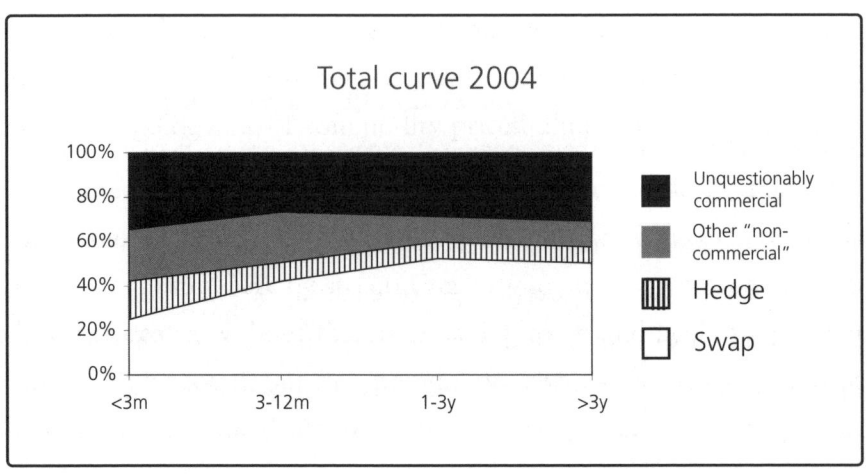

Total curve 2000

Total curve 2004

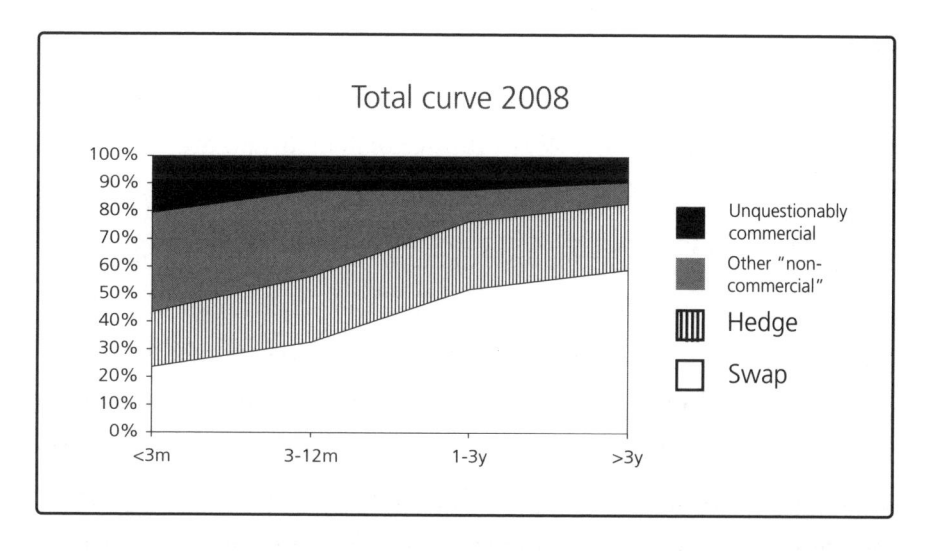

Figure 9: Changing trader group domination across maturities from 2000-2008
[Source: CFTC (Büyüksahin, Haigh, Harris, Overdahl & Robe, 2008)]

Compared to 2000, when the "unquestionably commercial" element of the market did hold at least half the contracts with maturities of under three months, and a slightly lesser proportion of three-year-plus maturities, by 2008 the unquestionably commercial market segment is clearly holding just a small minority of interest across all contract maturities, from front and nearby months right through to the longest-dated.

4.4 Mad May

It seems clear that the now-notorious Goldman note issued in mid-May 2008 would itself have actually contributed to swap dealer-transmitted financial speculator interest at exactly the farther reaches of the curve in particular. As a piece of research written by investment bank analysts for institutional clients, by definition it obviously serves speculative financial interest in the oil price. And the house recommendation the analysts made to readers on the basis of their reading of the market was indeed a long-dated single commodity swap on oil – namely buying the four-year forward Nymex WTI oil price, calendar-dated 2012, which was then sitting at $119.38 per barrel. This is the trade that institutional clients heeding the advice would place with a swap dealer, who would in turn then be seeking, all else being equal, to hedge this exposure by going long of the 2012 oil price on the actual Nymex futures market.

In fact, some commentators at the time would hold that Goldman note responsible for kicking off a particularly manic couple of weeks in the Nymex oil futures market – which can be summarised in terms of sequential futures curves, as shown in Figure 10. For on May 16 2008, the day the note was released, the whole curve shifted upwards

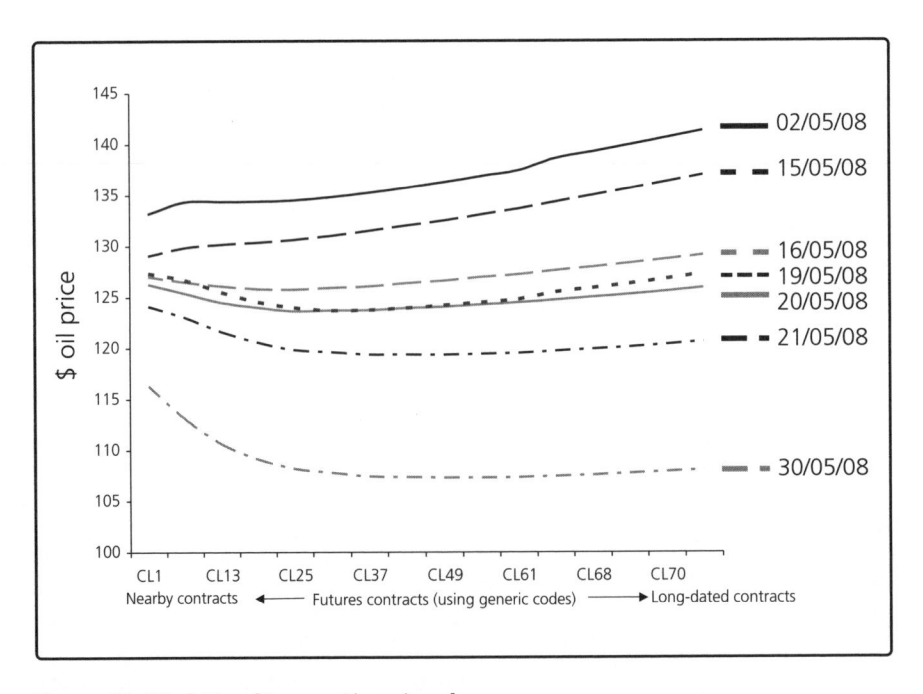

Figure 10: Mad May [Source: Bloomberg]

noticeably, but with the back-end gaining more than the front. While front-month crude jumped some $2 per barrel, the four-year (2012) price targeted specifically as a long trade by Goldman jumped almost $5 and the farthest-traded date, June 2017, jumped *more* than $5. This was to set in train a burst of price appreciation across the futures strip, characterised by the back end appreciating more than the front. Just two trading days after the Goldman research update, this trading activity had already caused the very unusual shift in the shape of the curve into a perfect, full contango, as referred to previously.

Compared to prices on May 2 (the first trading day of that month) at the most extreme point of this movement on May 21 the front-month price had gained 14.5%, almost $17 per barrel, the four-year (2012) price had gained 27%, almost $29 per barrel, and the nine-year (2017) price had gained just over $33 per barrel, or 31%. Then, as abruptly as it had appeared, this perfect contango collapsed back into the backwardation that had characterised the curve at the start of the month, but at a considerably higher price level. This can be seen in the curve for the last trading day of that month, May 30, at which point the front-month price closed up 9% compared to May 2, the four-year price finished up 16%, and the nine-year price up 18%. So while the curve had reverted into backwardation, still the long-dated break-out had left its mark in terms of which maturities gained the most through this month.

This rapid flexing of the oil futures curve from one shape to another and back again within days was unusual enough to draw analyst comment at the time. As Dr Ed Morse of Lehman Brothers commented in a research note released a fortnight after the Goldman note, 'Fundamental changes cannot explain sudden, severe price or curve movements.' Morse had his own explanation for the near-unprecedented price movements that week, which can be partially gleaned from the title attached to this research – "Oil dotcom". The front-page of the note made the point with a graph illustrating that the ascent of the Nymex oil price in terms of proportion over time mirrored the previous expansion of one of our more recent speculative bubbles, the ascent of the Nasdaq tech-stock market up to its peak in March 2000. The note argued, 'As in the dotcom period, when "new economy" stocks became popular, a growing number of Wall Street analysts have been repeatedly raising their forecasts as oil prices have risen. These revised forecasts have been partially responsible for new investor flows, driving prompt and forward prices to perhaps unsustainable levels.'

4.5 "Oil dotcom"

There were good reasons why the oil market should have paid attention to Ed Morse – his career spans decades in the industry across a variety of prominent roles. Apart from teaching international monetary policy at Princeton University, Morse had been Deputy Assistant Secretary of State for International Energy Policy in the Carter administration, had co-founded high profile energy consultants PFC Energy, and been publisher of oil sector trade news mainstay *Petroleum Intelligence Weekly*. After this he spent seven years as a strategic advisor at energy trader Hess Energy Trading Co (HETCO), and was then appointed head of commodity research at investment bank Lehman Brothers in 2006. Yet in May 2008 Morse found himself among a few lonely contrarians, as he warned the rest of the market that prevailing oil prices were a bubble waiting to burst.

The May 29 2008 Lehman research note "Oil dotcom" followed closely on the heels of another released on May 16, ironically the same day as Goldman's now-notorious $141 second half 2008 average oil price prediction. Entitled "Is it a bubble?", this had introduced some calculations Morse's team had been making regarding the weight that long interest index investment was bringing to bear on the Nymex oil futures market and hence the oil price. Morse and his team reckoned that commodity index investment alone accounted for as much as 25% of total open interest in Nymex oil by early 2008, and had even calculated that for every fresh $100 million inflow into commodity indices as an asset class, the WTI oil price gained 1.6%. The analysts also asserted that in the first quarter of 2008, this index investment had indeed played a key role in leading oil prices higher.

"Oil dotcom" built on this work but also noted how index investors alone were not the whole story, as the mix of speculative investment pressures on the Nymex curve grew richer in the second quarter of 2008. The note laid out in considerable detail how, through the remarkable weeks of May 2008 in which the curve flexed so dramatically, the whole oil futures curve was under bombardment along its length by a diverse artillery of speculative investment strategies. Alongside supposedly long-term, long-only index investors at the front end of the curve, Morse and his team identified short-term investors in long-dated options as the key source of buying pressure pushing up the long-dated end of the curve. It was this that was helping it flip into that relatively rare "perfect contango" we have already discussed – Morse and his team called this the 'anomalous deferred [i.e. long-dated] oil price move'. Because actual physical producers had notably failed to appear as sellers into this long-dated long interest, the long-dated end of the curve had shot up due to unbalanced pricing pressure. As the note said: 'In the past, producer selling would have slowed or impeded the upward price drive. In its absence, retail investors, hearing a growing number of reports about "peak oil", continued to buy at higher and higher strike prices, indicating that there is no necessary physical ceiling to the market.' In other words, there was nothing stopping option speculators from driving the long-dated price sky-high.

We should note at this point that this is only one of three different versions of what happened along the futures curve in those frantic May weeks in 2008. We have already heard the analyst anecdote concerning time spread investors having to unwind a trade going against them; then we have the account from Morse and his team highlighting short-term options traders *in particular* as responsible for flipping the curve so violently; and I have personally also been vouchsafed a version in which this unusual movement in the futures curve was due to a large commercial producer having to unwind some hedges that were going badly. Which is true? Perhaps all of them, to a certain degree. If the commercial producer unwinding its hedging position is doing so precisely because speculative forces have driven the oil price far above levels he previously thought suitable to lock in sales at, is his action a commercial decision or one forced by speculation? The point is that in any attempt to determine what was moving the oil futures market through the mad weeks of May '08, speculative investment patterns have to figure high on the list of factors for anyone claiming to listen to those who have tales to tell of those weeks.

The Lehman note did not dismiss potentially genuine fears about underlying supply and demand, such as the apparent diesel squeeze which exercised diesel fundamentalists like Philip Verleger and Francisco Blanch of Merrill Lynch. There is, however, an alternative account of this diesel squeeze which, while acknowledging real world facts lying behind the wide diesel crack spread, nevertheless reframes these facts as ephemeral and short-term considerations which should have been recognised and discounted as such by the markets, and therefore removed from consideration as fundamentals. The answer, as with so much nowadays, lies in China.

4.6 The Dragon's Hoard

China came out with a double-whammy on diesel demand for summer 2008. Firstly, a devastating earthquake in the Sichuan province on May 12 meant a very large-scale rescue and relief operation that was hungry for diesel. The quake also damaged power generation and coal transport infrastructure in that region, throwing many companies onto emergency power generators also fuelled by diesel. Secondly, the country was hosting the Olympic Games in June, and as part of its efforts to showcase China at this historic moment the government had established a policy of building up diesel stocks. This was both to ensure unforeseen supply disruptions did not affect the smooth running of the sporting festival; as well as to provide a substitute fuel for the coal normally burned in Beijing power stations through the weeks leading up to the Olympiad, in the interests of cleaning-up the capital's notoriously polluted air for the benefit of tourists.

This Chinese hoarding arguably led to an *apparent* shortage of diesel around the world, which in turn blew out middle distillate crack spreads. And the word "apparent" is justified because, with this information as part of the picture, it becomes clear that on an underlying, fundamental basis there was in mid-2008 sufficient light sweet crude being produced and likewise sufficient diesel being produced in the normal run of things. It was instead a large quantity of this existing and normally readily-available diesel being pulled off-market by the Chinese that made matters seem otherwise. Importantly, this Chinese diesel stock build would not have appeared on the oil inventory datasets normally perused by oil analysts. This sort of report, produced weekly and monthly by respectively the EIA and IEA, only covers the OECD grouping of which China is not a member. So the normal inventory figures that do the job well enough most of the time were blind to this particular instance of hoarding.

Yet it should have been recognised for what it was by the wider oil futures market – a temporary and unusual blip in diesel demand and stock-building which would be unwound as soon as the Sichuan rescue effort itself abated and the last bunting was taken down in post-Olympics Beijing. As such, this was clearly not a persisting, fundamental condition that spelled a permanent squeeze on light sweet crude and diesel output alike, and so should not have been seen as justification for the whole oil futures curve moving up to the heights it reached in summer 2008. That the market might be misreading this situation was highlighted by Dr Morse and his team in "Oil dotcom", which warned 'there is a tendency for the market to conflate legitimate reasons for demand growth with what is likely a temporary spurt in recent demand for inventory related to the Olympics. Once these are concluded and economics trump air pollution concerns, many Chinese thermal [power] plants are likely to switch back from diesel to coal.'

Morse and his team at Lehman Brothers meanwhile drew a stark conclusion from their own up-close, detailed analysis of who was buying what on the oil futures market: 'Our conclusion from this study is that we are seeing the classic ingredients of an asset bubble. Financial investors tend to "herd" and chase past performance, comforted by the growing analytical conclusion that markets are tightening, and new inflows, in turn, drive prices higher. Larger allocations by institutional investors, including new sovereign wealth funds desiring to increase their commodity exposure, play a role. So does uncertainty about the true state of market fundamentals, including the level of Saudi spare capacity, the level of Chinese "real demand" versus stockpiling, and other factors that bolster the current bullish consensus.'

The analytical standoff between an established body of opinion that triple-digit oil prices were justified, and a much smaller group of hold-outs convinced that prevailing crude prices were driven by speculators, continued as oil shot toward its peak. But the establishment could point to the big guns arrayed on its side. In early July '08, the IEA issued its own supposedly authoritative view on medium-term oil market fundamentals, in its annual *Medium-Term Oil Market Report* (MTOMR). The 2008 issue mounted a defence of fundamentals as responsible for prevailing oil prices, then fast approaching $150 per barrel, and concluded: 'While recognizing that speculation can have a day-to-day impact on price moves, the fact that all producers are working virtually flat out and that there is no sign of any abnormal stock build gives a strong indication that current oil prices are justified by fundamentals... Often it is a case of political expediency to find a scapegoat for higher prices rather than undertake serious analysis or perhaps confront difficult decisions.' The last dismissive comment summed up the consensus view that those obsessed with a speculative bubble in the oil markets simply did not understand how these markets work.

The initial results of the CFTC special call to swap dealers would not be publicised until the autumn. The *Interim Report* of the US government interagency task force was released a lot sooner, on July 22. In itself, this haste was a matter of some controversy – not least for one of the CFTC's own commissioners involved in preparing it, who as we shall see issued a dissenting opinion on the findings and the way they were prepared. Nevertheless, the official line taken in this report was that, 'to this point of the examination, the evidence supports the position that changes in fundamental factors provide the best explanation for the recent crude oil price increases. Observed increases in the speculative activity and the number of traders in the crude oil futures market do not appear to have systematically affected prices.' But by the time it was publicised, however hurriedly, this finding looked decidedly behind the curve. For the oil price was collapsing, in exactly the fashion which implied that $100-plus prices had indeed simply been a speculative bubble all along. We can say this, because we can compare the oil price blow-out to previous speculative bubbles.

5

A Bubble by Any Other Name

'When peak prices hit, we believe they are also likely to fall precipitously. That's the way cyclical turning points tend to occur – in the midst of a market trend, turning points can be sudden, unexpected, and severe.'

Edward Morse et al, Lehman Brothers, "Oil dotcom", 29/05/08

5.1 The Bust

One year on from writing the words above, Ed Morse confesses that he did enjoy some *schadenfreude* when what he had warned of for so long came to pass. But who wouldn't – because when the bust in the oil price anticipated by Morse, Colin Smith of Dresdner Kleinwort, and a few other market commentators (myself included), did come, it was truly spectacular. From oil's highest ever closing price of $145.18 on July 14 2008, a consecutive run of lower closing prices saw front-month Nymex oil finish that Friday July 18 2008 on $128.88 per barrel. Despite a few weak counter-trend rallies, it was really downhill all the way from there. By August 5 2008 the oil price had dropped through $120 per barrel, and on September 15 a 5% drop in one day saw it plunge through the $100 per barrel mark and close at $95.71.

This watershed must have shocked oil bulls into one final rage against the dying of the light, as the crude price then staged its most notable rally in this period, shooting back up to trade as high, intra-day, as $130 on September 22. But it was all for naught. Another drop followed, and by October 1 of 2008 the oil price was trading firmly back below $100 per barrel, a level which it has still not to date ever regained. From closing at $98.53 on October 1, by October 31 2008 the oil price closed at $67.81, losing 31% in a single month. By December 1, it had dropped through $50 per barrel, but even this was not the end of the retreat. Intraday trading on December 19 2008 saw the price drop as low as $32.40 per barrel – well under a quarter of its peak less than six months earlier. On the last trading day of 2008, the year which had opened with oil first touching $100 per barrel, Nymex front-month crude closed at $44.60 per barrel. Incredibly, oil had fallen so far it was back to where it had been four years earlier.

What had happened to $100 oil? The bullish case for triple-digit prices that had been pushed by many investment banks and organs of establishment opinion was simply shredded. In the face of such an unprecedented collapse in the oil price, many scrambled for new explanations that might explain the latest developments. For instance, by the end of that first week of declines ending July 18 2008, some market commentaries in the media were pointing to recent bearish updates from OPEC regarding the likely level of oil demand in the coming year as a cause for the nascent sell-off. This is, however, disingenuous – as we have seen, throughout the whole first half of 2008 OPEC had been warning that oil supply was exceeding demand and that inventories were building as a result. So the impression that the cartel had suddenly changed its tune and introduced new information into the market that traders were not previously aware of, and which might therefore justify an abrupt reversal in pricing, is completely misleading.

By the end of that first trading week after oil's peak, running from July 14-18 2008, oil had traded down 12% but no one could really point to any single, concrete fact to trigger such a pullback. As opposed to citing the OPEC outlook update, the *Financial Times* concluded on July 18: 'Oil had seemed immune to talk of weaker growth. But a downbeat assessment by Ben Bernanke, the Federal Reserve chairman, on Tuesday, and hopes of easing tension between the US and Iran, a big oil producer, appear to have at least temporarily taken the wind out of prices.' A completely different set of factors! And note how tentatively these linkages are asserted; the use of the words 'appear' and 'temporarily'. These factors were certainly not taken as conclusive signals that the oil price boom had ended, and the same report went on to note that 'Other traders were still cautious about calling the top of the oil market, having made calls in the past three months after similar price plunges only to see oil prices march higher later.'

Between July 14 2008, when oil finished the day's trade at what still remains its highest-ever closing price of $145.18, and September 16 2008 when oil closed at $91.15, oil shed 37% in two months. Just one day prior to that latter date, US investment bank Lehman Brothers filed for bankruptcy, setting in motion the train of events that saw the credit crunch grow to its most monstrous proportions, as banks simply ceased lending to anyone. Subsequently, many apologists for triple-digit oil forecasts say it was clearly the Lehman bankruptcy that "changed everything", something they could never have foreseen. But it is obvious that a massive first leg of the oil price collapse had already occurred before this point, more than a third of the peak price having already vanished.

On September 15 2008, the day the US government introduced a previously unthinkable scenario as it allowed Lehman to implode, oil had indeed closed below $100 for the first time since March 5 of that year, dropping 5% on the day. But two weeks earlier on September 1, oil had dropped even more, 6% in a day, with no obvious single news trigger. Energy risk management firm Cameron Hanover was quoted on that September 1 price drop at the time: 'The best reasons we can give for [it] are the strength of the US dollar, the continuing decline in consumer demand, and the market's recent trend lower. The reaction is telling us that this market just does not have the stomach it once did for higher prices.' Lehman Brothers collapsed, immediately darkening the economic outlook as it did so, well after the oil market had already turned. And as with other, previous bubbles, the exact reason for the sudden failure of momentum in the oil price right at the top of its peak, and its abrupt reversal into a steep plunge, remains unclear. Certainly, there was no sudden new nugget of information revealed to the market that might justifiably trigger such panic.

In fact, once the rout was underway the market actually ignored big news stories that might plausibly support the oil price or at least cushion its fall. The six-day war that erupted between Russia and Georgia less than a month after oil's peak was an obvious example, exposing grievous miscalculations not only by the Georgian president but also by the analysts who had claimed fundamentals were completely supporting $140-plus oil. In late June, the temporary loss of some 250,000 barrels of daily oil production (bpd) due to the Bonga shutdown in Nigeria was cited by many as sufficient reason for oil smashing through $140, even as the unilateral Saudi production increases totalling 500,000bpd failed to have any impact on price expectations. In contrast early August saw Russian military action plausibly threatening the Georgian section of the transnational Baku-Tbilisi-Ceyhan (BTC) pipeline, which can ship up to 1mbpd of Caspian oil to Western markets, while separatist guerrillas in Turkey bombed the Turkish section of the same pipeline, forcing temporary shutdown in any case. Yet through these same days, oil dropped another 4% to $113. Oil was dropping despite one of the most significant real threats to global supply to have actually emerged through the whole year. The implication that the price levels oil was relinquishing had in any case borne scant relation to reality, 1mbpd supply threat or not, was stark. Fundamentals do not evaporate in a month, but speculation does.

This radical disconnect from reality mirrored as much in the price deflation as in its previous appreciation is in itself also a hint that we are indeed dealing with a speculative bubble. When the market has already lost touch with real-world fundamentals, sometimes plausible grounds for the inevitable reversal seem as elusive as those claimed for the preceding inflation. Robert Shiller has searched in vain for actual real world news events that might have triggered two notable US stock market collapses, the Great Crash of 1929 and the crash of 1987. Shiller is instead forced to focus on the mechanics of what he terms a 'negative bubble', the equally irrational inverse of the previous manic price appreciation with exactly the same sort of feedback loop from price to sentiment and back to price again, but this time driving stunning price drops rather than gains. As he concludes, 'There is no way that the events of the stock market crash of 1929 can be considered a response to any real news stories. We see instead a negative bubble, operating through feedback effects of price changes, and an attention cascade, with a series of heightened public fixations on the market. This sequence of events appears to be fundamentally no different from those of other market debacles – including the notorious crash of 1987'[3].

The oil price crash of late 2008 seems to have followed the same pattern. Certainly through the initial phases, until the Lehman collapse gave grateful financial hacks a catch-all reason for every market setback across any asset class, the price drops in oil were themselves the main story, and the explanations proffered for this ongoing pullback were a hazy grab-bag of various observations. Interviewed in late May 2009, Jeffrey Currie of Goldman Sachs remains convinced that it was the Lehman bankruptcy which bankrupted in turn all his previous assumptions regarding how oil prices would play out through 2008 and

---

[3] Shiller (2005), pp. 97-98.

into 2009 – in all their triple-digit glory. But of the oil price drops from the peak until the Lehman collapse, he nevertheless concedes: 'The drop from $147 to $115 per barrel, that was idiosyncratic to the oil market – prices had gotten too high too fast, and there was also some hoarding going on.' In other words, people had simply been paying too much for oil compared to what was justified by fundamentals.

It is certainly true, however, that following the Lehman implosion, the oil price dropped at an even more rapid pace, until on December 19 2008 it traded as low as $32.40 and closed at $33.87 – the level which turns out to have marked the lowest point in the oil price collapse, prices having recovered since. From peak to trough the oil price saw an incredible drop of 77% in just under six months. We have already noted the lack of obvious news flow to trigger either the collapse from peak pricing in the oil market or the historic stock market collapses examined by Robert Shiller. In seeking support for the theory that the oil price boom-and-bust was a classic speculative bubble, we can point to far more coincidences than just that, however. The oil price boom-and-bust of 2008 displays a wide range of markers for the particular pathology that economists and historians have identified in previous episodes of speculative excess.

5.2 Forever Blowing Bubbles

Instinctively, anyone who has studied or worked in investment markets thinks they know what a market "bubble" is: a bout of stupendous price inflation in a particular asset class that seems to feed on itself, displaying a simple schematic of yesterday's price rises justifying purchases today which will again push prices up to justify yet more purchases tomorrow. But the concept of the "speculative bubble" in asset or market prices has only regained intellectual respectability in recent years, and in particular in the aftermath of the credit crunch and the associated collapse of the global financial system. Such a "bubble" can be defined as inflation in the price of a given asset to levels unsustainable in terms of an "underlying" or "real" valuation, and therefore by implication stoked by "speculators"; defined in turn as investors lured into an asset class for no more compelling reason than the hope of making money simply by running with the prevailing tide of investment fashion. But defined as such, the very existence of bubbles has been downplayed or relegated to the sidelines in the last couple of decades, a period notable rather for being characterised by the resurgence of free market fundamentalism.

The precepts of "efficient markets" and "rational economic agents" prevailing in the halls of economic academia and the brain trusts of the all-conquering global investment banks meant bubbles were a theoretical impossibility. Under such models, the price reached in a free market for an asset at any given point in time must perfectly reflect all the known information relevant to that asset price and is therefore identical with its "real" valuation, there being by definition no "underlying" valuation to reference beyond this from which market value can be stretched due to unwarranted speculative attention. Meanwhile in the citadels of central banking, from where interest rate pronouncements can actually have a definite impact in dampening financial sector excesses, any policy geared towards identifying or controlling asset price bubbles was effectively dismissed as a practical impossibility.

None other than long-time former US Federal Reserve chair Alan Greenspan, for many years the world's most powerful central banker by dint of his control over US dollar interest rates, had publicly declared in 1999 that 'Bubbles generally are perceptible only after the fact. To spot a bubble in advance requires a judgment that hundreds of thousands of informed investors have it all wrong. Betting against markets is usually precarious at best.' The message was, in other words, that while investment bubbles may occasionally actually exist – and note here the reluctance of an efficient market acolyte to gainsay the judgment of the legions of 'informed investors' in making such a determination – to all intents and purposes there is nothing to be done about them, other than ruefully recognise them only in the aftermath of their bursting. In regulatory as well as theoretical terms, then, bubbles were until recently essentially invisible, the ghosts at the financial feast.

Of course, ex-Fed chair Greenspan has plenty of scope for rueful recognition now he bears the dubious honour of asset price bubbles actually being named after him by a slew of commentators. The questions are no longer over the existence of these bubbles or whether central bankers should act to identify and pre-empt them (in the new, post-crunch financial orthodoxy – yes, they should), but over which particular bubble inflated under his monetary policy oversight is being referred to by use of his moniker. Is it the dotcom technology stock bubble that burst just after the turn of the century, or the US housing bubble that followed on the heels of the dotcom collapse, or indeed one over-arching bubble combining these two asset blow-outs, in recognition of the fact that Greenspan's persistently low interest rate medicine for nursing financial investors through the bloody aftermath of the former was also the tonic on which they bulked up to ridiculous extremes in the latter?

The world banking system has endured disastrous capital impairment, following the exposure of a gaping chasm between its own valuation of US property assets and what emerged as realisable proceeds from these assets. Since then, only a scattering of die-hard efficient market fundamentalists (admittedly including much of economic academia) would still contest the proposition that asset price bubbles are a real phenomenon, which in some essential sense involve the divergence of market valuations from what turn out to be the sustainable, long-run valuation of those assets. Unfortunately, it does indeed seem to have taken the crippling of the global economy to finally ram home to society at large the point that, where asset price bubbles are blown, they turn out to embody the very opposite of the optimal capital allocation that the efficient market Taliban claim for unfettered markets. But of course, in the neo-classical economic theory which denies legitimate usage for the term "bubble", the capital write-downs resulting from such asset price deflation are simply the "creative destruction" inherent in the market finding a new price equilibrium level due to new information becoming available, with this new price level neither no more nor no less "correct" than that which preceded it.

Back in the real world, however, when the gentle push from the invisible hand of the market becomes a hard shove in the back of bankers towards the edge of a valuation precipice, they are suddenly reluctant to accept that what the market will pay for their assets at any given time is necessarily the correct one. Sometimes the market valuation is "distorted", or "out of line with fundamentals", argue the bankers – as they ask taxpayers to take toxic assets off their hands at higher prices than open sales would realise. Equally importantly, politicians who formerly competed for the accolade of being the most "pro business" in their embrace of free markets are no longer so keen to agree that any pricing equilibrium reached by these free markets at any given time must be the correct and only one possible, when the pricing equilibrium now implied across almost every asset class by markets left to run their course unimpeded spells significant job losses and prolonged economic hardship for wide swathes of their electorate.

Economic policy wonks have suddenly rediscovered the wisdom of British economist John Maynard Keynes. His arguments included the proposition that there are in fact multiple equilibria possible for markets, that some equilibria are qualitatively better than others when judged in terms of efficient employment of all capital available (including the supply of human capital), and that sometimes external intervention is required to ensure that these are the points at which markets settle. Such notions formed the basis for the long postwar boom up to the 1970s, but were studiously disregarded by most pundits courting economic credibility from the 1980s onwards. But look around the world now at the panoply of soft loans, subsidies, capital injections, and other forms of taxpayer-funded state aid being enacted to "artificially" stimulate economic demand – and so resuscitate ailing banks and industries (state bounties awarded to private citizens for new automobile purchases being the latest wheeze). It is clear that we are all Keynesians once more.

Rediscovered alongside Keynes is the American economist Hyman Minsky, himself an intellectual disciple of Keynes at a distance, and often painted as a fellow-traveller in Keynesian economic heresy against efficient market theory. And so he is. Rather than agreeing that markets tend toward settling around a stable equilibrium at the price that most efficiently matches supply of a particular asset to demand at any particular moment, his "financial instability theory" postulates instead that markets are perennially unstable. Rather than each new price level or supposed market equilibrium always being a logical reaction to a change in external stimuli – in other words new information being made available to the market – price levels also evolve in relation to what might be termed "feedback effects" inherent in the workings of the market itself. These feedback effects are the prime motive forces behind market bubbles.

5.3 Minsky's Moment

Minsky's feedback effects are essentially the same phenomena alluded to by Robert Shiller in his work already quoted. In gaining a fuller picture of how these effects arise and are transmitted through the market, Shiller's account of feedback in *Irrational Exuberance* can be set profitably alongside previous work by Minsky himself (for example, in *Stabilizing an Unstable Economy*), and the discussion by George Cooper of Minsky and instability theory in his recent book *The Origin of Financial Crises*. Synthesising these approaches into a single chain of arguments, we can begin by saying that the aforementioned feedback effects are manifested through diverse elements of market structure such as the fractional reserve lending model of banking, or the discipline of mark-to-market accounting used by lenders to control the leverage they make available to investing institutions. These feedback loops have an amplificatory effect such that an initial externally-triggered move in the value of assets can cause waves of secondary adjustment throughout a market and build up to events many orders of magnitude greater than apparently foreshadowed by the initial stimulus.

For example, losses announced by a bank on a particular loan, or indeed even suspected by informed and risk-averse savers, can cause a withdrawal of funds by some depositors because the liability representing the cash claims of all depositors at that bank is now backed by a relatively more impaired or overstretched income-generating asset base. Yet in withdrawing their own cash from the bank's capital base, itself already just the small, liquid fraction of that asset base held close to hand by the bank, they unavoidably put the equivalent right of instant cash withdrawal for the remaining depositors at greater risk than it was theoretically put at due to the original loss on the loan. Yet more depositors realise this further heightening of risk and withdraw their own cash, putting the claims of still-remaining depositors at even more risk. And because everyone knows that under conventional fractional reserve lending a bank only has cash equivalent to a tenth or less of its actual deposited liabilities on hand, soon depositors are flocking to withdraw their cash and there is a run on the bank.

Or in the case of mark-to-market accounting, a mark-to-market loss recognised by an investing institution on an asset which was purchased with debt leads to a "margin call" by its lender for cash to cover the difference between the debt extended and the new and now-lower realisable value of the asset. To raise the cash required to either cover the margin call or pay back the debt in full, the institution has to liquidate some of its other positions, in other words sell assets other than the one on which the original loss has been recognised. But selling these other assets increases the downward pressure exerted in reaching the market equilibrium that determines their price at any given time, and if this change in pressure is in turn enough to lower the publicly recognised market value of these other assets then the institutions that remain holding such positions have to realise a fresh round of marked-to-market valuation adjustments. If they are also leveraged with regard to these positions, they will in turn also face a fresh round of margin calls – and so on, such that forced selling to meet margin calls on losses on one asset class can inevitably spiral outwards into a systemic cascade of market value write-downs across a wide range of assets.

As described above, the feedback effect that swamps the market is initially triggered by a supposedly external stimulus. It is nevertheless Minsky's particular genius to demonstrate that with these feedback effects structurally built into the market, there is a natural tendency toward instability regardless of external stimuli. Or, as Minsky himself says, thanks to 'disequilibriating forces that are internal to the economy... success in operating the economy can only be transitory; *instability is an inherent and inescapable flaw of capitalism*'[4] [his italics].

As he shows in *Stabilizing an Unstable Economy*, for example, it is in fact long periods of apparent market stability that lay the groundwork for a subsequent boom and collapse. As returns achieve an ever-longer record of stability and therefore apparent robustness, growing confidence in this robustness swells market appetite for risk such that lower and lower cushions of safety are built into the financing models used by investors in assessing investment in a particular asset and the financial institutions that lend to them. Lending structures evolve from what Minsky labels *hedge* finance, through *speculative* finance, and into *Ponzi* finance (the latter, ultimate stage of development bearing the same name derived from a world-famous "pyramid scheme" peddler that Shiller later uses in his own explanation of naturally-occurring bubbles).

In hedge finance, it is presumed that successful repayment of the debt in question is possible simply on the basis of the future cashflows to be obtained from the asset it is funding, and the quantity of debt offered is based on prudent and conservative estimates of these likely future cashflows. In speculative finance there is an acknowledgement that cashflows attached to the asset may not be after all be sufficient to repay existing debt as already agreed at some points in time, but also a presumption that if such occasions do indeed arise, refinancing will be available in the market because the eventual value of all receipts will indeed support this level of debt, only now paid back over a longer period. In Ponzi finance, any semblance of safety in the financing structure is foregone, with the simple presumptions both that the debt assumed will definitely need to be rolled over due to inadequacy of near-

---

[4] Minsky, Hyman P., *Stabilizing An Unstable Economy*, (McGraw-Hil, 2008), p134.

term receipts from the asset, and also that refinancing will nevertheless be available at the requisite moment because the asset price in question will meanwhile have appreciated in value.

In the end it seems inevitable that the Ponzi financing will evolve to a stage where even the slightest perturbation in underlying asset price – even the most minimal shift generated by a random walk model of market price movement – will be sufficient at some point to render the terms of the next pending refinancing finally and fatally unattractive to those as-yet uninvested in the asset. Ponzi financing cannot bear the first hint of strain on its structural presumptions. As with a pyramid scheme, the first gap in the virtuous circle of investment driving prices higher and so itself encouraging profit expectations in the next wave of investors quickly expands into a yawning chasm. Suddenly there are no takers for the asset at ever-higher prices – that perennial supply of "greater fools" has at last dried up. At this point all the feedback mechanisms that have inflated the bubble kick into reverse, fuelling not rapid price appreciation but precipitous price decline. Thus stability has finally bred its own destruction, regardless of any required critical mass of external stimuli. Depending on the particular stage in market evolution from hedge, through speculative, to Ponzi financing, the same asset price move that might spin the latter into crisis might not disturb a hedge financing structure at all. But given the tendency of all markets to gravitate toward Ponzi financing, instances of explosive asset price appreciation and equally rapid price contraction are a question of when, not if.

We have of course seen these sorts of negative feedback effects throughout the unfolding of the credit crunch, with crises such as the run by depositors on the Northern Rock mortgage lending bank in the UK which led to its state takeover (the first such bank run in the country for more than a century). Another example would be the mass deleveraging by margin call-plagued hedge funds from late 2007 onwards, which saw the prices of supposedly "good" assets take a near-indiscriminate hammering alongside obviously "toxic" assets. This in particular revealed just how much of the preceding generalised asset price boom was dependent on ever higher and higher levels of leverage, ever increasing quantities of borrowed money, and ever tighter margins of viability on that debt compared to the range of outcomes reality might throw at the presumptions underpinning the financing. The last days of the housing boom that afflicted so many developed world economies (particularly the US, UK, Spain, and Ireland), and brought us the sub-prime mortgage debacle, was a prime example of Ponzi finance at work.

The lesson from Minsky and contemporary proponents of his theories such as George Cooper is that asset markets do not simply stabilise themselves at an equilibrium appropriate to the external environment, but rather have an inbuilt structural tendency to inflate beyond the pricing levels fundamentally appropriate to that environment, even in the absence of significant external stimuli. The result of the overshoot can be an asset price bubble which, crucially, contains within itself the seeds of its own reversal – and on the way back down, the structurally-determined overshoot can tend towards recession or even the Keynesian "liquidity trap" of outright depression. Crucially, overshoot in either direction can continue for a very long time, but nevertheless, due to the unstable foundation on which it is ultimately based, it is liable to reverse at practically any point and will with time eventually inevitably collapse into its opposite trend.

These periodic reversals from market expansion to contraction or vice versa are those "Minsky moments" you may well have been hearing so much about in recent media discourse. For example, in the expansionary bubble scenario that is our main concern, a continual ratcheting-up of market values beyond the appropriate valuation trend actually justified by each positive piece of news increases the chances that the next time a piece of negative news comes along it may well expose the magnitude of this widening mismatch. As seen, it is in fact a symptom of bubbles that the next piece of bad news signalling this mismatch may well be ignored, and so indeed may the one after that and so on. But there always remains world enough and time for a particular piece of bad news to eventually be noted and acted upon by a particular market participant, and then the snowballing feedback effect has an opportunity to take hold in the reverse direction. When it does, that is the Minsky moment.

5.4 The Pathology of Bubbles

While recognising the force of these theories allows us to say asset price bubbles are real as opposed to simply a label for market developments we happen to be caught on the wrong side of, nevertheless we are left with the problem that bedevilled Alan Greenspan. As George Cooper describes, bubbles are very hard to definitively label as such in explicitly economic terms whilst they are being inflated. This is precisely because the feedback effects which amplify market movement into unwarranted price inflation or deflation are transmitted from one key indicator to another to create a largely convincing picture of a market in optimal equilibrium, rather than precarious instability constantly teetering on the possibility of trend reversal.

In the case of mark-to-market accounting, gains in the market valuation of assets posted by investing institutions as collateral against loans used to buy those same or other assets seemed to justify further extensions of credit by lenders to these investment institutions. In turn, these fresh infusions of borrowed money into the asset market raised the upward pressure on price-setting equilibria, leading to higher market valuations and another increase in borrowing capacity. Each expansion in leverage is theoretically caused only by an appropriate increase in notional asset values, so if you were trying to spot a bubble prior to it bursting by measuring growing investor overexposure in terms of debt to asset value, you could well be wasting your time. After the bubble bursting, of course, the newly-revealed value of the assets does indeed look far too small to support the leverage extended on their basis – but, by then, it is too late.

Feedback transmission channels are not, however, limited to the sort of debt leverage "force multipliers" of marked-to-market asset collateral or margin calls. Robert Shiller also identifies prevailing media discourses as a major reinforcer of irrationally bullish market sentiment. These discourses relate to each price movement through an explanatory prism of pre-conceived and largely untested notions regarding how the particular market is going to develop in the future, which Shiller labels as 'New Era' thinking. An example would be the dotcom era ideology that real world profits no longer mattered as long as you could show enough sales growth to justify an ever-higher stock market valuation, which would in turn allow you to raise the next round of equity needed to tide the company over until real profitability finally appeared – if it ever did.

Psychologically, the constant repetition in the media of such overarching narratives creates an environment in which lenders get more and more comfortable with the idea that the apparent equilibrium exhibited in the expansionary trend is indeed deeply entrenched, so they worry less about potential risk and relax their standards. In short, they get sloppy. Would that we had the correct instrumentation at our disposal, the degree of this lender sloppiness is no doubt as revealing with regard to the length of time for which a particular investment trend has been flavour of the month as the half-life of carbon-14 is with regard to the age of dead organic matter.

Nevertheless, the tendency for bubbles to inflate through parallel and mutually reinforcing market indicators means that, in spotting bubbles in advance, we must rely on less purely theoretical but more behavioural or pathological patterns. In this respect we can quote a useful summary from veteran financial journalist Robert Teitelman:

> Bubbles are assets that are inflated to such a degree that we heave the anchor on intrinsic value and sail merrily into speculative fantasy. Asset bubbles show a precipitous spike as the mania takes hold. Asset bubbles are driven higher as speculative types pay attention only to each other, not to the world outside. Asset prices spiral. Bubbles typically display a feverish, frantic, divorced-from-reality mentality that increases as time moves on – until the sudden break and collapse.

Teitelman goes on to distinguish the concept of a "bubble" from far more generic miscalculations or misreadings of markets sometimes labelled as such, mistakes of the sort that wreck countless real businesses in the day-to-day vicissitudes of enterprise but which in themselves don't justify the label. 'A bubble,' he argues, 'is a deeper, narrower phenomenon that represents not the foul-ups of senior managers but the descent of an entire market into a self-fulfilling and self-reinforcing daydream. The dotcoms were part of a bubble; so were sub-prime mortgages.' Writing in May 2008, Teitelman was referring to the two most recent obvious bubbles at that time, but, in fact, another one was being inflated in the oil markets, even as he penned these words.

The oil price spike of 2008 displays all of the key markers of the pathology of bubbles described above. Feedback loops are present in force. Examples include the tendency of large chunks of long-only, passive index investment in the OTC derivative market to add upward pressure to prices on the Nymex market itself; thereby seeming to justify fresh waves of new investment into the index structures, which then again need to be matched by long hedges on the Nymex, once more raising prices. Or within the Nymex exchange itself, the fact that investment at one portion of the futures curve can drag the price in another portion of the curve up or down, which may trigger yet more latent stop orders beyond the original point of price action, which lead to their own rallies in turn.

The feedback loops inherent in the dynamics of leverage and mark-to-market accounting, explored so thoroughly by Minsky and Cooper, are of course central to the very functioning of a futures market which settles trades financially on margin accounts. Obvious examples of the sort of 'New Era' thinking described by Shiller include the "commodity super-cycle" presumptions and the peak oil narrative. We have already seen how the media word count of several such memes increased through the first half of 2008. Through this period the oil price bubble clearly displayed many moments when price developments were obviously at odds with a fundamental account of underlying supply and demand – or as Teitelman's pathological focus might put it, investors in the oil price were not paying attention to the world outside and were divorced from reality, venturing into a self-fulfilling, self-reinforcing daydream.

With regard to how the pricing trend reversed, there are indications that it could simply have been due to the eventual exhaustion of investment appetite for ever-higher oil prices. As suggested in the newspaper report above – which referenced investor 'stomach' for ever-higher oil prices – what might have happened, in other words, was precisely the sort of exhaustion of "greater fools" that signals the collapse of a Ponzi financing scheme, rather than any noteworthy real-world development. Ironically, Ed Morse and his team at Lehman lost their jobs right in the middle of the oil price crash that vindicated their whole stance on the market, as their bank collapsed in its own sub-prime mortgage Ponzi financing meltdown. Very soon afterwards, Dr Morse and Lehman colleague Daniel Ahn were employed anew by broker LCM Commodities, under which aegis they have since published research indicating that from the end of May 2008 onwards there was indeed a series of weeks seeing significant net withdrawals of index investment from the futures market, which Ed Morse himself puts down to 'rebalancing'. For instance, see LCM Commodities research note entitled "South by Southeast" dated March 19 2009 for a graph showing estimated crude oil futures index investment-related flows, clearly showing that the massive outflows which came to characterise the second half of 2008 actually started in earnest from June 2008 onwards, in other words before the crude oil price had actually peaked.

Rebalancing is an idea we shall examine presently, but this evidence essentially reinforces the picture that just after the madness of May, there was indeed a cooling of investment ardour towards the oil price. That notional next wave of new investors perhaps did not quite match up to what had gone before in terms of appetite, and from that point on the oil price was wobbling. It probably did not take much to bid it up further to $147, yet by the same token it probably did not take much either to tip it into the "negative bubble" that was always going to end this expansion. This feeds into another marker of speculative bubble pathology – the fact that, under the instability theories described above, the market correction to the bubble in question is as likely to overshoot to the downside as the bubble has overshot on the way up, particularly given the leveraged, mark-to-market structures much financial speculation is funnelled through.

When the oil price collapsed, it did not collapse back to levels which many $100-plus sceptics themselves would have thought reasonable – perhaps the $60-80 per barrel range – but instead dropped past these prices by a long way. That the price sunk to levels far below reasonable can be seen in the fact that oil gained some $10 per barrel in exiting its trough between Christmas 2008 and the start of 2009: roughly a 30% gain in two weeks, from prices which have not since been revisited. The $34 per barrel oil was fetching around Christmas was clearly as unrealistic as the $147 it had fetched in early July. But speculative bubbles may, after all, overshoot on the way down as well as on the way up. And in regard to the severe rebalancing, manifested in this sudden lack of appetite, there remains another loose end to tie-up.

Readers may have wondered why more has not been made of the graph presented earlier in the book, which plotted the growth of the oil price alongside the explosion in Nymex open interest from 2000 through to the end of 2008 (Figure 5). Clearly the period that saw a rapid oil price ramp-up also saw a rapid ramp-up in open interest in the Nymex oil futures curve. Why do we not simply say that here is yet more proof that the increase in paper barrels was driving the oil price? In truth, some have. Roger Diwan of PFC Energy, for one, calculated an R-squared of 81.5% for the statistical measure of correlation between movement in the oil price and the growth of open interest on Nymex in 2003-2006. In other words, 81.5% of the movement in the oil price could be "explained", in statistical terms, by movement in Nymex open interest through this period.

There are problems, however, with asserting this kind of simplistic relationship between the oil price and reported Nymex open interest. For a start, Diwan also found a much lower R-squared, of just 30.8%, emerging for correlation between open interest and the oil price across

2006-2008. While both indeed eventually did end up some 50% higher at the end of this period than at its start, the noticeable drop in the oil price toward the end of 2006, even as open interest rose, shreds the correlation across this whole period. Beyond this observation, in statistical jargon there is the general problem of calculating an R-squared across two trending sets of data – i.e. two sets of data where there may be an inbuilt bias toward expansion regardless of their relationship to each other (for example, growth in trade on a market from its inception, or growth in a commodity price that will necessarily trend upward over time with inflation). My friend and econometrically-learned colleague Chris Dillow, for example, notes how a high R-squared can be calculated between growing US national debt in the 1980s and deaths from AIDS through the same period – but are we really meant to believe this proves some sort of causal link between the two?

We cannot simply say that rising open interest equals rising oil prices, and vice versa. Sometimes a significant linkage may appear, other times it will not. And this should be obvious, given something we have already touched on earlier – the additive rules of open interest. The prices of futures are determined by the movement of trading volumes at various points along the curve – which, as we have seen, is not the same as the open interest left extant after this trading. And this is in turn a key point to bear in mind when considering an objection sometimes made against assertions that speculative interest blew the oil price bubble. This line of argument runs thus: if a massive weight of speculative interest piling into oil futures supposedly caused prices to rocket, why then when the oil price later collapsed did the exit of all these speculative players not make more impact on open interest? As Figure 5 shows, this actually stayed roughly level or even grew at times as prices tumbled. It is the additive rules of open interest at work again. Even if there was a frantic sell-off on the part of speculative investors working to push prices down through the negative bubble effect, open interest would only decline if the same players who originally sold future oil at triple-digit-plus prices to these speculators were liquidating their positions at the same time. But why would they, when plunging prices meant these positions were getting more valuable every moment? Open interest can remain the same, even as previous long speculators herd out of the oil price en masse, if the market participants buying them out of their positions are not the original sellers also offsetting, but new players tempted into these positions by the latest price drop. And indeed, newspaper market reports from the period support this view with anecdotal evidence, noting how, in the early weeks of crude's plunge, sellers of oil at high prices were content to sit on growing gains, forcing long liquidation to pass through a narrower channel of available buying interest than would otherwise be the case. All of which, of course, would work only to increase the price drops with each round of liquidation.

5.5 Inside the Bubble

The oil price blow-out of summer 2008 undoubtedly *looks* like a speculative bubble from many angles. We see the price behaviour exhibited over time uncannily matching that witnessed in the notorious dotcom blow-out. We see a patent lack of support from physical market fundamentals, which should have been apparent from early in 2008. We see a well-documented explosion of speculative financial investor interest in commodities in general and the oil price in particular in recent years. We see obvious mechanisms to transmit this generally long-only interest directly to the futures market-determined global benchmark oil price itself; and a wealth of "feedback loop" amplification effects, both psychological and structural, obviously reinforcing this momentum. And lastly we see the resulting price overshoot of the Nymex oil market itself, both on the way up and the way down.

The arguments supporting the view that speculative activity was instrumental in oil prices peaking at over $140 per barrel in July 2008 should now be clear. Which class of market participant drove the speculative bubble in oil prices that peaked in summer 2008 – the hedge funds simply looking to maximise short term returns, the "dumb money" piling into long-only commodity index investing as a long-term proposition, or more sophisticated investors likewise exerting influence via the OTC hedge but nevertheless exploring deeper into the maturity curve? The correct answer is, without doubt, all of the above.

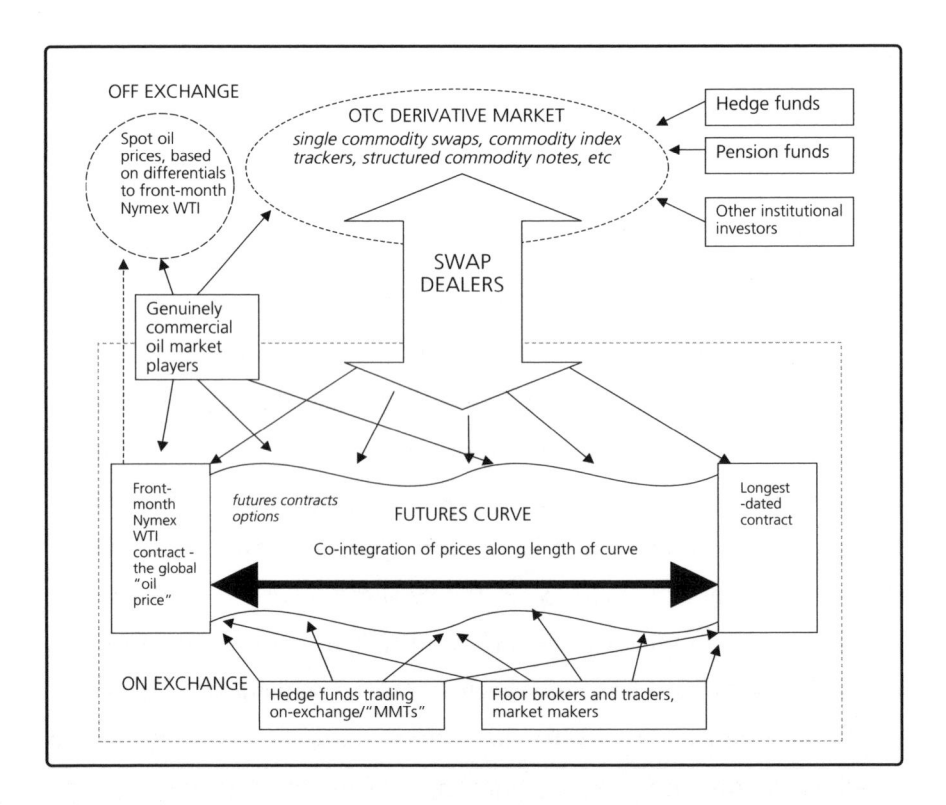

Figure 11: Forces bearing on oil price formation in the Nymex futures market

Figure 11 is a simple schematic visualising the wide spread of speculative financing interest feeding into price determination alongside genuine physical market considerations.

Yet any account focusing on speculative interest as the main driver behind the oil price in summer 2008 must answer some questions of its own, and address the issue of how to square this conclusion with evidence offered to the contrary, from a variety of sources, which supports rival accounts. The two major points that must be dealt with and fitted into any account of the price blow-out which seeks to hold water are:

- Firstly, the diesel fundamentalist argument asserting that it was a genuine and fundamentals-based squeeze on the light sweet segment of the crude market which ramped prices up. This view must be taken seriously because of the indubitable blow-out in middle distillate crack spreads over summer 2008, which accompanied the more obvious crude price spike; a pairing which would indeed normally indicate fundamental shortage in the market.

- Secondly, the assertion that academic studies, particularly those under the auspices of the CFTC with its enhanced access to Nymex trading data, prove that speculative interest does not materially affect price formation in the oil futures market. This view must be taken seriously simply because it is still repeated by many, even after the US regulator itself has had to do some serious trimming and tacking with regard to its own previous position on the extent of speculative trading in the oil futures market.

A key plank of the most convincing argument that oil prices at last year's peak were indeed justified on fundamental grounds is the evidence from the middle distillate crack spread that there was a shortage of diesel fuel at the time. Advocates of this stance argue that – as there was no discernible middle distillate stock build in the figures analysts had access to at the time – this was a genuine shortage. As such, the obvious cause is a genuine shortage of light sweet crude to produce diesel from.

But we have already seen how Chinese hoarding due to earthquake relief and Olympic preparations is a very plausible candidate for *artificially* giving the appearance of an undersupplied diesel market in May and June 2008. So yes, for some weeks through this period at least, the market was receiving signals which seemed to indicate a fundamental shortage of diesel and therefore of light sweet crude. But no, this apparent shortage was not fundamental at all and had, as with any other apparent shortage not based on fundamentals, its own accompanying stock build somewhere around the world, only this one did not show up clearly on the usual radar because of its Chinese provenance.

This Chinese stock-build, however, became more noticeable to many as it was unwound following the Olympics and pressure on the middle distillate segment of the market was relieved, with a notable drop in crack spreads from June 2008 into autumn. Interviewed in May 2009, even Jeff Currie of Goldman Sachs acknowledged China as a factor in 'some hoarding' he now admits was indeed going on through the oil price peak. Although he is reluctant to give it a starring role, Currie says of Chinese diesel hoarding in particular: 'That was going on, but at the time we thought it was pretty small. Now we have data to look back on, we can see that it wasn't *big*, but it was bigger than we thought.' Other oil sector experts such as Ed Morse, Leo Drollas of the Centre for Global Energy Studies (CGES), and Stephen Schork, however, disagree. As far as they are concerned, the Chinese hoarding did indeed make a significant difference to the wider market perception of middle distillate market balance, whereas it should have been recognised as a temporary aberration.

Moreover, the whole thesis of a diesel squeeze justifying a structural, long-term shift in light sweet crude oil demand, and therefore the global benchmark oil price, fails in the face of another key consideration. Everyone knows there is a whole swathe of new refining capacity coming onstream in the next year or two, concentrated particularly in China and India, which is specifically configured to crack diesel out of heavier, sourer crudes. It is this prospect, coupled with the reduction in global demand we are now experiencing, but which was also clearly emerging through 2008, that has now seen middle distillate crack spreads recently mark fresh lows since their peak last summer.

Overall, it has to be conceded that the particular conditions of diesel demand on the ground for a couple of months in summer 2008 were real enough, and that in this diesel squeeze we can also see evidential justification for Dr Philip Verleger's concern about the US government hoarding of light sweet crude in its Strategic Petroleum Reserve. But it must also be recognised that this diesel squeeze was a brief and passing phenomenon. While on fundamental terms it might justifiably have added some dollars to the oil price through May or June 2008, it did not justify the oil price getting to the levels it had already reached by that point. The diesel squeeze was essentially a thin layer of fundamental icing on a large speculative cake being baked in the Nymex oil market in early summer 2008, and it by no means rebuts the case that speculative finance was the major driving force in reaching triple-digit oil.

A priori, any statistical study attempting to measure speculative influence on the oil price by using simply the traditional CFTC split between its own definitions of "commercial" and "non-commercial" market participants is invalid – simply and obviously because of the swap dealer loophole that allows speculative financial interest in the oil price to be hedged on Nymex under the "commercial" cover. At a stroke this rules out a whole slew of previous studies, including those pre-2008 studies conducted under CFTC auspices which were repeatedly referred to by regulatory officials like Walt Lukken in rejecting the suggestion that financial speculation was driving high prices. But don't just take my word for it – no less august a body than the United States' own Government Accountability Office (GAO) reached much the same conclusion in a briefing entitled *Issues Involving the Use of the Futures Markets to Invest in Commodity Indexes* prepared for the US Congress and released at the start of 2009.

Commenting on a selection of studies it had reviewed, the briefing notes: 'Four of the studies used CFTC's publicly available Commitments of Traders (COT) data in their analysis, and their findings should not be viewed as definitive because of limitations in that data… these data generally aggregate positions held by different groups of traders and, thus, do not allow the effect of individual trader group positions on prices to be assessed.' The same briefing also goes on to mention two other reports prepared with non-public CFTC data. These are the same two reports the CFTC released from summer into autumn 2008 – the *Interim Report on Crude Oil* released in July '08, conducted by the Interagency Task Force on Commodity Markets set up and chaired by the CFTC, and the *Staff Report on Commodity Swap Dealers & Index Traders*, which the CFTC released in September.

The GAO briefing itself notes that as these two studies found no evidence of speculative trading materially affecting oil prices, this would be its own provisional view as well. The findings of these two studies themselves are, however, open to question on several grounds.

5.6 What the Studies Do and Do Not Say

The *Interim Report on Crude Oil*

The *Interim Report on Crude Oil* released by the Interagency Task Force on July 22 2008 actually marked the first emergence of detailed data on the Nymex trading positions of specific types of market participant that would subsequently appear in a CFTC report released in December 2008. This latter CFTC report, entitled *Fundamentals, Trader Activity and Derivative Pricing* and authored by Büyüksahin, Haigh, Harris, Overdahl and Robe, is itself not directly concerned with the effect of speculative investment on the oil price. It is instead an updated version of the February 2007 CFTC study already mentioned (and written by the latter four of the five authors above), which established that there was increasing co-integration between the nearby and farther-dated portions of the futures curve and fingered swap dealer activity as the major cause of this phenomenon.

Whereas the February 2007 co-integration study compared 2000 and 2006, the dataset as presented in the late 2008 update is split between 2000, 2004, and 2008, and is the basis of the figures presented earlier in this book to show the true extent of potential speculative interest on Nymex compared to what is implied by the traditional CFTC distinction between commercial and non-commercial. The Interagency Task Force indicates in a footnote that it has taken its data from the then-pending 2008 update on the co-integration study. However, as it presents figures for every year from 2003 to 2008, the underlying dataset must be more than the 2000-2004-2008 snapshots presented in the 2008 co-integration update as released. Regardless, the Task Force took this data and then analysed it in a different way to reach its own verdict on the matter of whether or not speculative interest materially shaped oil prices on Nymex.

In line with previous CFTC pronouncements, the Task Force found that 'the activity of market participants often described as "speculators" has not resulted in systematic changes in price' – so no surprise there. Yet despite itself, this report still contained some eye-opening information for those wondering at the scale of potential speculative weight on the market. For whatever reason, the Interagency Task Force did not actually present the data in terms of tabulated hard numbers (as the update to the co-integration study would do in December 2008). Instead it produced blocky, 3D-effect bar charts to show open interest for each category of market participant, very hard to measure against the sparsely scaled and strangely offset vertical axis it also used. As a result, scrutinising the report graphics, it is impossible to say exactly where in the space between 750,000 contracts and 1 million contracts the average open interest on the market held by swap dealers in the first half of 2008 falls. It looks, however, to be somewhere around 900,000 (as already seen, the table in the subsequent co-integration update, used for the charts earlier in this book, puts the figure at 947,952).

Even the unhelpful presentation format used in the Task Force report could not, however, hide the fact that swap dealers held three times the volume of contracts held by the next-largest "commercial" market participant, the physical oil traders, with actual oil producers and fuel refiners negligible in comparison. And it was also clear that adding in another 750,000 or so contracts from the hedge funds and floor traders detailed in a separate graphic for "non-commercial" participants meant the total potential speculative weight on the market in early 2008 appeared to be near enough seven times that of the unquestionably commercial market participants. (In actuality, the same 85/15 split seen in the chart earlier in this book.) And newspaper reports following on from the release of the report would indeed focus on how speculators appeared to hold 80% of the market, far more than previously presumed.

As well as confirming the predominance of swap dealers, the Task Force report also revealed that since 2004, the net exposure of all swap dealers trading on Nymex had been on the long side, i.e. as a distinct group of market participants they were net buyers of oil – although this net weight on the long side had declined markedly since 2006. Despite all this, however, the Task Force report said that it had performed batteries of statistical tests to establish whether or not trading by these different classes of market participant could be seen to cause movement in oil prices. Its conclusion was that neither the officially non-commercial players (including hedge funds), nor the swap dealers considered as a group in themselves, could be seen as leading price changes by their own positioning of volumes in the market, but that these traders followed on the heels of movement by genuinely commercial market players.

But this was qualified: 'While these statistical tests present the most complete examination to date of the relation between position changes and price changes, they – like all statistical tests – have some limitations. First, the analysis was performed for trader groups rather than individual traders. Consequently, these tests would not be able to detect if the positions of some traders within a trading category have much greater influence over prices than the positions of other traders in that category. Second, the tests utilise end of day position data. Thus, the tests may not capture any intraday position-price relationships. Finally, the tests were performed on aggregated net position changes in the nearby contracts alone (defining nearby as the futures contract with the largest open interest). As a result, the tests do not reflect a systematic effect of position changes at different maturities on either the prices of the nearby futures contract or on the whole term structure of futures prices.'

We could therefore pick holes in the Task Force conclusions on all of the grounds above; these indeed being real issues, any one of which might upset the tentative conclusion. But perhaps the admitted focus that these causality tests had on the front end of the curve is the most striking, given the evidence already suggesting that nearby prices can move in sympathy with longer-dated prices. It therefore could well be actually along the further reaches of the curve that a particular instance of pricing pressure in any direction from speculative market participants originates, rather than right at the front end of the curve, where instead the price might be moved by more commercial players *responding* to what they see going on further out the curve. Even on its own terms, then, the Interagency Task Force *Interim Report on Crude Oil* from July 2008 leaves no little room for legitimate questioning of the conclusion it drew.

The CFTC *Staff Report*

Compared to the two months the Interagency Task Force had from its inception by the CFTC to its first and thus far only report being issued, the in-house CFTC *Staff Report on Commodity Swap Dealers & Index Traders* (the *Staff Report*) which was released in September 2008 had a significantly longer gestation period. (Not least because it was to be an opinion reached on data that had only been requested in May 2008.) Strikingly, however, it is quite hard to actually determine any definite conclusion made regarding the effects of speculation. It is presented as a summation of data received from the "special call" for swap dealer trading information, and then a list of recommendations regarding how the CFTC should modify its regular data collection and presentations, without much significant reference to whether or not the information unearthed in the "special call" proved or disproved arguments regarding speculative pressure in the market.

Yet the *Staff Report* is nevertheless cited by some, including the GAO, as helping establish that speculative pressure does not lead oil pricing on Nymex. Why? The answer is in its presentation of one particular segment of data, returned by the special call, regarding the volume of long exposure to the oil price it calculates as being held by swap dealers due to commodity index investment. The report finds that while the net notional value of long-only commodity index investment in crude oil futures rose from $39 billion at the start of 2008, to $51 billion by end-June 2008, this was due simply to the value-at-stake in contracts rising as the oil price itself rose. However, in terms of volume of contracts bought in expectation of the oil price rising, the *Staff Report* finds that long positions associated with commodity index investment actually declined through the first half of 2008.

In other words, some measure of commodity index investment was actually being reversed out of Nymex even as the oil price was soaring toward its peak. The implication is therefore that commodity index investment cannot be responsible for the oil price breakout if it declines through the same period. Herein is the basis on which this particular report is seen to deny the speculative case. Dealing with this inference requires a closer look at the data – which says that while the oil price was $96 per barrel on December 31 2007, as index investment accounting for the long side of 408,000 Nymex contracts, by March 31 2008 the oil price was $102 per barrel but long index interest would have been some 398,000 contracts; and by June 30 2008 the oil price was $140 per barrel while long index interest would have been 363,000 contracts.

It is true, then, that as the report states, 'Measured in standardised futures contract equivalents, the aggregate long positions of commodity index participants in Nymex crude oil declined by approximately 45,000 contracts during this 6 month period – from approximately

408,000 contracts on December 31 2007, to approximately 363,000 contracts on June 30 2008. This amounts to approximately an 11% decline.' Yet there is another way of putting this. That 11% decline was not evenly spread over the six months in question, but concentrated from March 31 2008 onwards. Up to that date, only 10,000 contracts were liquidated, just over 2% of the supposed index interest from end-December 2007.

It is actually between end-March and end-June 2008 that the rest, almost four-fifths, of the 11% decline comes, with 35,000 contracts being liquidated. This timing is important, because when seen as a phenomenon overwhelmingly concentrated in the second rather than the first quarter of 2008, this reduction in long index investment is exactly what you would expect to see on the part of speculative investors profiting on a roaring oil price – it is evidence of "rebalancing". Rebalancing is a very simple concept and practically universal across professional investors. It simply means that when an investment is particularly successful, over time an investor will trim their position in it both to realise profits and also control risks.

For an extremely simplified example, imagine you had put $1000 into oil at the end of December 2007. By the end of March 2008, with oil having gained 6.25% from $96 to $102 per barrel, let's unrealistically imagine (i.e. without transaction costs or other forms of dissipative "friction"!) that our own investment has gained the same, and is now $1062.5. So far, so good – our investment has risen in value. Not so spectacularly, however, that we feel obliged to do anything other than let it run on. By the end of June 2008, however, oil is $140 per barrel, now up 40% year-to-date. Likewise, our original $1000 is now $1400 – which is great, but presents its own, albeit agreeable, "problem". This problem relates to exposure to a single asset, in this case oil. After all, it is not just our profit that has grown 40%, but also our exposure to oil itself. Considering we were originally happy with just $1000 of exposure to oil, perhaps we are now concerned with leaving too much capital, almost half our original stake again, at risk to future oil price performance.

Rebalancing is the answer. Selling out of some of the notably successful investment realises profits now, and importantly for fund managers and other institutional investors also locks in at least some measure of positive performance on the investment to date, while also reducing the exposure to the particular asset back towards where it started out originally. Whether or not the rebalancing goes right back to the original level of investment, and at what levels of gain and over what timeframe it is carried out, are all real world considerations with an infinity of permutations depending on who the investor is. But it is easy enough to see that while there would be no great incentive for investors in oil to rebalance a position held over end-December 2007 to end-March 2008, by the time the end of the second quarter approached in June '08 there would be a lot of incentive for rebalancing.

Looking at the dates and figures quoted in the *Staff Report*, there is no further information to be had between the end of March and the end of June 2008. We simply don't know at which point in this timeframe the bulk of the index position liquidation in question occurred. So it would be as valid a statement as that made in the report (namely, that in the six months leading up to the peak of the oil price, long index positions on the Nymex exchange declined 11%), to say instead that the evidence equally plausibly supports another interpretation. It might instead back up the view that index investors could have held onto some 98% of their end-December 2007 positions right up until the last weeks of June, only then liquidating another 9%

of end-December holdings just in time to lock in profits for first half reporting, and only after oil was close to $140.

In behavioural terms, this is exactly what might be expected of such investors. And there is a big difference between this latter scenario and the tenor of the *Staff Report*. In the first, these liquidating index investors are still in the game until late June, when there is only a fortnight and some $7 to go before oil reaches its peak; but the implication in the *Staff Report* wording is that these index investors were systematically disinvesting throughout the first half of 2008, and were therefore unlikely drivers of oil towards its peak. As with so much else in this story, the truth no doubt lies somewhere in between – but again, as with so much else in this story, with the information available at this point we simply do not know for sure. Dr Ed Morse, for one, certainly feels that rebalancing by investors was evident after the remarkable weeks of May 2008, a month in which oil not only clocked up some of its strongest gains (in dollar terms) through 2008 but also punched through both the $120 and $130 per barrel thresholds for the first time.

5.7 The Dissenting Commissioner

Both the *Interim Report on Crude Oil* and the CFTC *Staff Report on Commodity Swap Dealers* make statements dismissive of the idea that speculative investment has a material role in determining oil prices which, upon examination, are questionable themselves. There is, however, another important consideration to bear in mind when weighing their contribution to the debate. This is simply that one of the CFTC's own commissioners involved in the preparation of both reports took the unprecedented step of issuing a public dissent from these findings, on the basis that he did not feel the data justified such a determination, as well as to express his concern at how such categorical statements were being made without what he regarded as sufficient evidential support.

Included as Appendix G to the *Staff Report* is the "Commissioner Bart Chilton Dissent", in which the eponymous CFTC staffer writes:

> I have significant concerns relating to the underlying analysis on which the recommendations are based. Specifically, I express doubts regarding the amount and type of data received in connection with the special call survey, the nature of the analysis, and I have a fundamental disagreement with certain conclusions underlying the Commission's recommendations. I am concerned that, while I believe the staff did a tremendous amount of work in a short period of time, the agency may not have received the type of comprehensive datasets needed to make reliable analyses and conclusions. It is my understanding that staff have been in the process of receiving data within just the last few weeks, making an extremely short turn-around time for processing and analyzing this exceedingly complex issue. Most importantly, I regret that certain conclusions may underlie these recommendations regarding the causality link between index traders and price movements, particularly in the crude oil market, and that these conclusions that do not appear necessarily to be ineluctably linked to the data received. Absent compelling evidence, I believe that the most responsible course of action is to refrain from making conclusions or declarative statements based upon such limited and unreliable data.

While Chilton's dissent is included as part of the *Staff Report*, in the same statement he also undermines the conclusions reached in the earlier *Interim Report on Crude Oil* released by the CFTC-directed Interagency Task Force, already discussed above:

> The Interagency Task Force released an Interim Report on July 22 [2008] that contained statements about speculative activity in the oil markets. Those statements were at best premature given that key information, such as the type of data on which the CFTC bases its instant recommendations, was not available for analysis by the Interagency Task Force. As noted, the reliability of the data and the conclusions of the staff's analysis cause concern that, should the Interagency Task Force Final Report rely on that analysis and review, its findings and conclusions will be similarly tainted.

In summary, then, none of the reports depended on by those claiming the CFTC proved that speculation does not play an important role in determining the oil price, have actually proved any such thing. Studies that rely simply on the traditional CFTC CoT distinction between "commercial" and "non-commercial" traders are disqualified by virtue of failing to pick up on the swap dealer loophole. Lest this ruling-out of much previous work conducted around this issue, particularly by the CFTC, be thought highly convenient for those pushing the argument that speculation looms large in the events of last year, it also rules out at least one other study which supports the idea that speculative investors can move oil prices, conducted by an academic outside the CFTC (see bibliography and source list for details). But sauce for the goose is sauce for the gander.

The two studies that move beyond the traditional CoT distinction to examine the trading patterns of more specific groups of market participants, including swap dealers, were both conducted under CFTC direction and both not only fail to convince in ruling-out speculation as a major influence on oil prices but are also undermined by one of the regulator's own officials taking the highly unusual step of drawing public attention to their multiple deficiencies. Meanwhile, neither the Interagency Task Force nor the CFTC itself have since released follow-up studies to either report. This is particularly striking in the case of the Interagency Task Force, which was meant to issue a final report in December 2008 but has thus far, by summer 2009, failed to come up with the goods. Whether or not the task force is still active is itself a moot point. No such follow-up report commitment was explicitly made by the CFTC with regard to the staff report on swap dealers, but it did say at the time that its new level of surveillance on swap dealers in particular would be maintained and that new ways of transmitting this information to the wider markets should be adopted – as yet, however, the form of the weekly CoT report remains stubbornly locked in the traditional outdated definitions of "commercial" and "non-commercial" participants which fox any accurate knowledge of speculative weight playing on the market.

The truth is that we have yet to have an authoritative, comprehensive study analysing the oil futures market in terms of the more detailed data categorisations the CFTC introduced in these two reports. Yet we should note that as more and more data emerged over summer 2008, the balance of probabilities seemed to be moving with those who saw speculation as a major driver of the market. Everything tended to support this: from the sheer scale of swap dealer predominance in the Nymex oil futures trading in general, to the fact that when heavyweight oil trader Vitol (see later) was reclassified from commercial to non-commercial on the basis that it was speculating, it was nevertheless found to be holding more than a tenth of open interest across the curve. As Commissioner Bart Chilton wrote in his dissent:

> In sum, smart people can disagree on this issue. Accordingly, until we have a comprehensive, unbiased study of this issue, we should not be making declarative judgments as to causation or effect. That being said, due to the critical nature of this issue, I believe the American public can no longer wait for studies or task forces. Given the data that we have received, I believe

there are some appropriate conclusions that can be drawn, and certain recommendations that should be made. I believe, first and foremost, there is a great deal of information regarding the over-the counter market that the agency does not routinely receive and that such information may be imprecise. This data may have a significant impact on the exchange-traded markets. Secondly, in my opinion it would also appear that such over-the counter activity may, at times, have uneconomic effects on prices.

5.8 Darker Portents

The documents described above certainly do not exhaust the list of academic, official, or quasi-official reports issued with regard to the oil price spike of 2008. Some were more qualitative in approach, as opposed to quantitative reports seeking measurable statistical evidence to support their statements. Among the former, for example, were the *Accidental Hunt Brothers* reports produced by Michael Masters (the first of which, at least, was produced at Congressional behest), which were in turn monstered so roundly by Dr Philip Verleger, again in more qualitative than quantitative style. Among the more intriguing overall, however, were the contributions from Robert McCullough Jr. of Portland, Oregon-based McCullough Research. These combine both quantitative and qualitative elements, and emerged from his testimony regarding the July 2008 oil price spike before the US Senate Committee on Energy and Natural Resources, on September 16 2008.

It is no surprise that McCullough was present before such a committee. In a long career he has worked, variously, as a professional economist, a university economics professor, and a long-time energy industry executive in roles spanning regulation, operations, finance, analysis, and strategic planning. As head of McCullough Research he has appeared before US Congressional and legal court hearings as an expert witness on energy markets – most notably, perhaps, in relation to detecting and describing the illegal market manipulation strategies used by disgraced and defunct energy company Enron.

It is in this latter regard that his contribution to the oil price spike debate stands out. As with many of the expert witnesses that testified on the oil price before US legislators in summer 2008, McCullough feels that fundamentals signally failed to justify the oil price. As he stated in his Senate testimony, 'A careful review of the fundamentals does not explain why the price of oil increased by 50% in the first six months of this year [2008] and then fell by 50% in the next three months. Supply and demand stayed in rough balance over the first nine months of 2008.' Yet his focus for an alternative explanation was a step beyond the "financial sector speculative pressure" theories as laid out in detail through this book. For McCullough found evidence that the same sort of pricing and trading patterns betraying the power market manipulation practised by Enron, at the start of this decade, could also be perceived in the Nymex oil futures market at the height of the summer 2008 price spike.

This assertion casts the whole debate about speculation versus fundamentals in a new light. Throughout this book, while the argument made has been that speculative investment took over from physical market fundamentals in determining the all-time high crude price of July 2008, there has been no suggestion of criminal behaviour. None, in fact, is necessarily needed. The attempt has been to prove that this is an all-too-obvious outcome of structural features across both the oil market itself and a financial institutional investment sector which comes to this market, in one way or another, seeking high returns. In this sense, despite the unfavourable outcomes attached to an oil price driven by speculators – such as galloping inflation, the failure of markets for bona fide physical hedging, and the immiseration of those suffering fuel poverty worldwide – there is no group of people to "blame" as individuals. If McCullough is correct, however, there may well be, after all, individuals directly culpable for illegally and covertly manipulating the price of oil futures.

The particular form of market manipulation that McCullough can arguably see evidence of in the oil futures markets over summer 2008 is referred to as the "spot forward gambit". Simply put, it involves a group of traders conspiring to profit from illegal collusion in trading by orchestrating simultaneous placements of large-volume buy orders in the spot market. This leads to a jump in price at the front end of the curve which, thanks to the phenomenon of "curve shift", the co-integration between nearby and long-dated price movements, moves prices up all along the curve. And by selling into this rally from positions held further along the curve, the colluding traders make more than they expend on bidding up the spot end of the curve in the first place. McCullough claims that the pattern of selling exhibited overall by "non-commercial" parties on the Nymex as the oil price moved toward its absolute peak echoed patterns exhibited by Enron traders as they moved to liquidate profits out of their own "spot forward gambits" years earlier.

Such tactics only work if the participants have sufficient market power to be what McCullough labels a 'pivotal trader' – that is, a trader whose market power is such that their own trading operations can swing the market one way or another regardless of market fundamentals. McCullough himself has no idea who might actually have been acting in such a fashion. His conclusions are based on studying the traditional CoT data recording all "non-commercial" market participants as a single block. But he does note that the reclassification by the CFTC of oil trader Vitol from its "commercial" into its "non-commercial" category in mid-July 2008 revealed the scale of market power that some individual traders do hold on Nymex. On July 15 2008, Vitol alone held more than 10% of positions (by volume) across the Nymex light sweet crude curve.

If McCullough is correct, then the oil price spike of 2008 may not have been just a matter of blind herding by speculative investors in expectation of ever-higher prices. To a certain extent it could also have been deliberately engineered by traders large enough to knowingly and deliberately move the market regardless of fundamentals. We should be careful, however, about attributing this exact synthesis of arguments to McCullough himself. As he wrote in email correspondence with this author, 'You will notice that I am on the milder range of the conspiracy theorists in 2008. We have a good set of data supporting oligopolistic behavior. We do not have the supporting data that would cross the line into conspiracy or deception. The self-serving forecasts of investment banks with large undisclosed positions are on the edge – a bit like crying fire in a crowded theatre and then rifling the possessions that the fleeing theatregoers left behind.'

McCullough notes that arguments attributing a key role to speculative financial sector interest in ramping-up prices around the end of June/early July '08 in particular must somehow explain a curious fact. The "risk premiums" embodied in long-dated prices along the forward curve – compared to existing oil price forecasts such as the EIA outlook – dropped through this period. In his view this is actually at odds with a picture of investors clamouring to get in at the long end of the curve (although, as noted above, there is also the possibility that financial speculators were indeed deleveraging/rebalancing exposures toward the height of the peak in any case). Nevertheless, he answers his own point by also noting that this line of argument presumes that such a risk premium behaves as if reflecting a fixed quantum of volume, to which sellers of future contract commitments will limit themselves. In fact, reality can be different: 'The unsophisticated version of this theory assumes that forward contracts are like beach property. If everyone wants beach front, the prices go up, since there is no more tomorrow than there is today. Forward contracts aren't limited, however. People will continue to sell you forward contracts as long as you have money – and then convene to pick over your bones after you run out of money.'

As for the general concept of a speculative bubble affecting oil prices through the period leading up to the all-time high in July, McCullough says in personal correspondence: 'I am certainly not one to ridicule bubbles and panics. We see them all the time. I am doubtful about the sheer scale of the panic last spring [2008] and its sudden deflation. I was curious enough last August [2008] that I ran the prices for any number of "announcement effects" and came up with virtually nothing. In Westerns, the cattle don't stampede until the cowboys fire their pistols. Last summer they apparently surged off on their own and then skidded to an abrupt stop.' And in the end, McCullough makes the same point in his September 2008 testimony made by many other expert witnesses through this tale: that the information at hand is woefully inadequate to categorically confirm what happened.

As this testimony stated: 'The events in the oil markets over the past nine months make it clear that none of these agencies or the nation's policy-makers currently have enough information to make informed decisions… The resemblance of the July 2008 oil price spike to earlier spot forward gambits is troubling. Even more troubling is that data on WTI crude spot and forward prices gathered by FERC, the FTC, the CFTC, and at the EIA is too insufficient to determine whether the price of oil was manipulated. Even more disturbing, last week's CFTC report [the *Staff Report* on swap dealers] that minimised the effects of speculation on oil prices, chose to stop its analysis in June, prior to the price spike.' In the battle of expert opinion regarding speculative pressure in the oil market, the CFTC has certainly not had the last word.

5.9 Diagnosis: petromania

In the final analysis, when speaking of visible proof of the speculative bubble, there is also the graphic trace the oil market left in its spectacular rise and fall, the chart of price movement over time. While Teitelman speaks of the 'precipitous spike' that marks a speculative bubble 'as the mania takes hold', Shiller meanwhile notes in his description of the S&P Composite index toward and then through the US market bubble of 2000 how 'the price index looks like a rocket taking off through the top of the chart, only to sputter and crash'[5]. Figure 12 shows the trajectory of the Nymex front-month oil price through 2008 in the visual context of a long-term pricing graph, running from April 1991 to the time of writing.

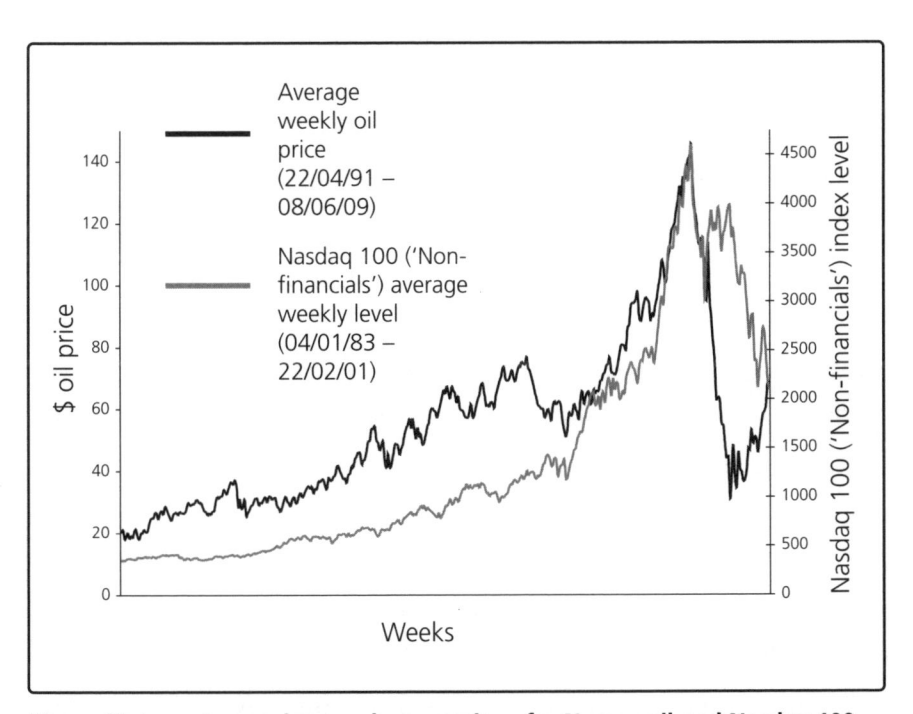

Figure 12: Long-term price graph comparison for Nymex oil and Nasdaq 100
[Source: Thomson Datastream]

---

[5] Shiller (2005), pp. 97-98.

The movement around the peak certainly does look like a 'precipitous spike' or an aborted rocket launch. Shown for comparison on the same graph is the previous trajectory of another price graph; if anything more closely tied to the turn-of-the-century "new economy"/dotcom/tech stock blow-out in the US than the S&P graph Shiller uses. This is the Nasdaq 100 ("non-financials") index, which practically embodies the dotcom boom-and-bust.

The Nasdaq 100 ("non-financials") index is comprised of industrial, technology, retail, telecommunication, biotechnology, health care, transportation, media, and service companies trading on the Nasdaq market in the US. It specifically excludes banking and other financial institutions listed on the Nasdaq, which from our point of view might add an unwelcome element of solidity to our sample of largely "weightless" (which turned out to mean profitless) bubble stocks. From trading at around 1,900 at the start of 1999, the index had reached its peak around 4,700 in March 2000 (in tandem with the wider Nasdaq composite index), at which point many of the firms included in the Nasdaq 100 were seen as the vanguard of the "new economy" focused around dotcom, new media and biotech narratives. The whole Nasdaq market collapsed spectacularly after this, such that by the start of March 2001, just a year after its peak, the Nasdaq 100 was back to around 1,900. The Nasdaq boom-to-bust is commonly seen as one of the more obvious examples of speculative excess blowing an unsustainable market bubble in recent years. Although some economists have nevertheless continued to hew to the precepts of efficient markets by trying to prove that at each point in its expansion and subsequent collapse the Nasdaq was indeed correctly priced in fundamental terms, such efforts are generally treated with gentle mockery.

Figure 12 shows how the pattern of both the Nasdaq 100 and the oil price blow-out sketch the same general shape when viewed from a wide

A Bubble by Any Other Name

angle. But, as shown in this book's introduction, the coincidence between the two price inflation and deflation patterns is actually far closer than just an apparent fit from a distance. In the introduction we saw how in three phases in particular, two leading up to the peak and then the third being the immediate post-peak collapse, the Nasdaq 100 and the 2008 oil price display practically follow the same pattern of percentage increases and decreases over the same number of weeks.

Figure 13 is an expanded version of the chart in the introduction, which also shows how despite divergence in their patterns following the three key phases above, nevertheless, after another 38 weeks showing broadly similar net declines, what is striking is that both price series end up being down near enough the same percentages from their absolute highs – the Nasdaq being 51% of its peak, oil 47% of its peak. Are we witnessing similar, post-bubble behavioural logic?

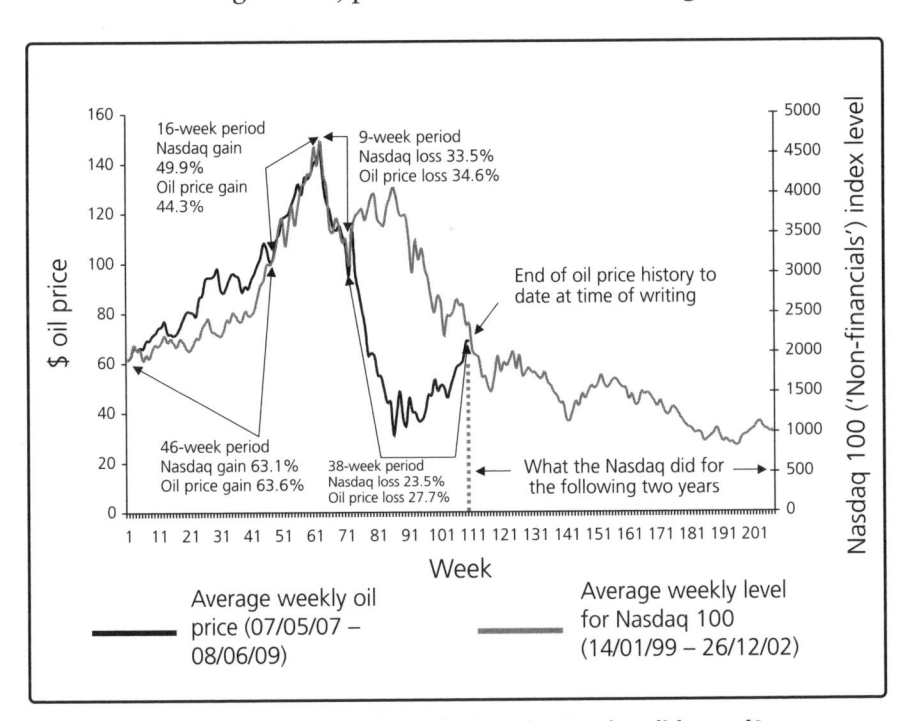

Figure 13: Two bubbles revisited – and what the Nasdaq did next [Source: Thomson Datastream]

For the oil price, this 38 week period in fact takes us right up to the time of writing, in early June 2009. Currently, the oil price remains less than half its all-time peak less than a year ago. Figure 12 also shows how the Nasdaq 100 index developed for the two years following its own peak period as already matched to the oil price – for no more reason than to show "what happened next". The next chapter will relate how even less than a year on from their peak, oil prices are once more displaying all the signs of another bubble being blown. So we cannot say if the most recent oil price levels displayed in the chart above (from June 2009) of around $70 per barrel, roughly matching the relative price level the Nasdaq displayed the same distance from its own peak, will turn out to be a near-term ceiling or not. Perhaps oil will bounce around from current levels – as indeed the Nasdaq 100 seemed to do for a while from this point on, in its own trajectory. Alternately, perhaps their superimposed histories will once again diverge – perhaps a fresh oil market bubble will indeed emerge regardless of how recently the previous one collapsed. That certainly seems to be a possibility at the time of writing.

6

Petromania Redux

'Like the toxic debt of the banks, this is a very esoteric subject. About a billion barrels per day are traded in crude futures on Nymex and ICE for WTI and Brent. A billion barrels. Then we have the options, and then we have the OTC market – which could be another billion barrels, but quite a lot of that will be covered by the swaps writers on the futures market. So avoiding double counting, we could take maybe 1.2-1.3bn barrels as a round number – a vast number. And this is the point I keep making, I keep saying here is a market that is fifteen times the actual crude market – and people just nod their heads and then they carry on talking about fundamentals.'

Leo Drollas, chief economist of the Centre for Global Energy Studies (CGES), in conversation, January 26 2009

6.1 Three Wise Men

As a working journalist I was fortunate enough to come down on the right side of the triple-digit oil price debate in early 2008. For many reasons, none particularly original to me and practically all of which are repeated at some point elsewhere in this book, I reckoned that triple-digit oil was a bubble being blown with a diverse spread of speculators working at the bellows. The Goldman research note in mid-May, followed by the remarkable fortnight of curve-shifting the oil futures market underwent immediately afterwards, stung me into printed comment in my day job for the *Investors Chronicle*. On May 29 2008 I argued that 'Goldman is wrong, and oil price highs above $100 reflect speculation more than reality'. By June 12 2008, personally shocked that prices were still breaking toward $140 per barrel, I advised readers to go short of the crude oil price by using a short exchange-traded fund (ETF) instrument listed on the London Stock Exchange (LSE), which would gain in value as the oil price dropped. This proved to be among our most profitable pieces of trading advice in 2008.

Contrary to the evidence you hold in your hands, I am, however, not a monomaniac regarding the oil price. My day-to-day brief is to cover the whole natural resources sector (mining as well as energy); there are other stories to follow and life goes on. Even so, when I attended an oil and gas investment conference in London in December 2008, with oil then bouncing around the $40 per barrel level, I felt it was so obvious that the whole $100-plus per barrel episode had been a speculative bubble that I naively presumed this would be acknowledged as a simple fact by practically any delegate speaking in the programme. I was surprised then, that in a "Three Wise Men"-type panel debate involving a trio of oil sector luminaries, the view that speculation had played any significant role at all in the amazing oil price blow-out seen earlier that year was very much a minority position – outnumbered two-to-one. The three wise men were Christof Ruehl, the chief economist of UK-listed international oil major BP, Francisco Blanch, head of commodity research at investment banking giant Merrill Lynch, and Leo Drollas, the deputy executive director and chief economist of the Centre for Global Energy Studies (CGES), a London-based, Gulf-funded energy consultancy.

I have great respect for both Christof Ruehl and Francisco Blanch, neither of whom I know in any meaningful sense but both of whom have been extremely helpful in commenting on aspects of the oil market whenever I have spoken to them in my work as a journalist. Both also know far more about the oil market than I ever will – so with these provisos firmly in the reader's mind, I will nevertheless take the liberty of characterising (but hopefully not caricaturing) the positions they adopted in this debate. For Christof Ruehl of BP, OPEC is the beginning and the end of the story. As far as he was concerned, the price ramp-up had been due to the OPEC cuts in 2007 feeding through into the market by early 2008, and then the unilateral Saudi production increases in turn feeding in to loosen the market just as the massive fall in global economic demand caused by the credit crunch hit in any case – an unfortunate coincidence. Ruehl said speculators of course participated in the oil futures market, but they waited to see where the price was heading before clambering aboard to make money in the wake of trends driven by physical market fundamentals, and OPEC is the single largest physical market fundamental in the game. Francisco Blanch meanwhile took the diesel fundamentalist line: the squeeze on the light sweet crude segment of the market was real, it was due to galloping Asian demand and if the credit crunch had not intervened, prices would have kept rising anyway until they themselves choked off sufficient demand to rebalance the market.

As his two co-panellists spoke before him, Leo Drollas looked increasingly perplexed. When he had the floor after them, he started by saying how amazed he was that in all the discussion that had already gone on, no one had mentioned what he labelled as 'the elephant in the room'. This proverbial pachyderm was the sheer size of the crude oil futures market compared to the underlying crude oil physical market and, as a result, the outsize influence that the futures market bore (and continues to bear) in determining the oil price. Drollas is himself a former head of energy studies and econometric analysis at BP, one of Ruehl's predecessors, so he has as good an understanding as any oil company professional of the fundamental factors at play in oil price determination. He was nevertheless adamant that until the effects of speculation were understood, it was futile to talk of fundamentals. But

Francisco Blanch himself responded that everyone could be sure that, precisely because the Nymex was such a deep, liquid and efficient market, speculation really was not an issue in pricing. In this dispute between our three wise men we have, in microcosm, the broad shape of the three key arguments running through this book, competing for our attention.

One is the background of physical fundamentals that oil company professionals like Christof Ruehl see in action every day. This is a world in which OPEC and its spare capacity, or lack of it, are considered the ultimate origin of all pricing shifts. Secondly, there is the tighter focus on Asian hunger for particular segments of the crude barrel refining split, and the squeeze this seemed to bring to the light sweet crude end of the oil market in mid-2008. This view is not overly adamant as to whether or not this light, sweet squeeze was caused by genuine market fundamentals or instead some kind of hoarding, deliberate or not, perhaps even shading into market manipulation. Thirdly, we have the focus on speculative pressure from financial investors as a key driver of the oil price. This driver has been the major concern of this book. Obviously, all three angles must be taken into consideration when rendering an account of the wild volatility our most important global commodity price demonstrated through the course of 2008. No reasonable version can omit any of them. Yet this is what many "establishment" accounts persist in doing even after the oil price collapse – as exemplified in the lonely stand taken by Drollas on the importance of speculation at this December 2008 conference.

Hence the focus throughout this book has been on forcing recognition of some simple truths regarding financial speculation, its impact on the oil futures market, and its major influence, therefore, on the oil price in the real world. All the requisite transmission mechanisms exist for a huge variety of financial investment strategies and instruments to exert pricing pressure on the all-important front-month price on the oil futures curve. Yes, of course physical supply and demand fundamentals had tightened through the course of the mid-noughties, creating the conditions for the oil price to establish a new floor significantly above $50 per barrel. And on top of this, geopolitical risk concerns from time to time were seen to add an additional premium on the per-barrel price of oil, a more or less reasonable outcome depending on your view of how likely events such as a fresh war in the Middle East were. But in my opinion it seems clear that speculative financial interest was the main force driving, and wildly driving, the oil price from its break through the watershed $100 per barrel mark at the start of 2008, to its as-yet all-time high of around $147 per barrel in early July 2008.

Physical market fundamentals do not support the breakneck price appreciation in front-month Nymex light sweet crude through this period. Neither at the broadest macro-economic level in terms of anticipated global growth, as the world was tipping into the serious recession we still find ourselves in; nor at the level of worldwide balance between oil production and consumption. Inventories were building, OPEC member states were claiming difficulties in selling all their production, and oil company economists and executives were confirming that the market was well-supplied. The marginal cost of production was below triple digits. Nor even after drilling down a further level into the particular drivers of the light sweet crude segment itself, do we find any significant enough support for these price rises. Chinese diesel hoarding was already recognised by some commentators as an ephemeral, short-term phenomenon that would evaporate in due course; and certainly not a long-term factor outweighing the imminent increase in sour-to-diesel refining capacity emerging in Asia. Inasmuch as the pressure on the light sweet segment of the market was real even if recognisably temporary, it should have been adding a premium to an oil price still comfortably trading in double-digit figures rather than to an oil price trading well into triple-digit territory.

Even physical market fundamentalists such as Christof Ruehl of BP can clearly see a disconnect between how the oil price behaved in the last months prior to it peaking near $150 per barrel in early July 2008, and how it logically should have behaved through this period. Speaking at a press conference in June 2009, presenting BP's latest *Statistical Review of World Energy*, Ruehl gave a potted review of the extreme volatility energy markets experienced in 2008. And when he mentioned the significant unilateral Saudi oil output increases of May and June that year, he himself noted, unprompted, how even though these developments should have driven the oil price lower, instead it jumped higher regardless. As an in-house oil company economist, concerned with taking the long view on whether oil prices averaged across several decades in the future will justify large capital investments set to produce crude through the same timescale, Ruehl may understandably not be so interested in an oil price obviously diverging from physical market fundamentals for, in his opinion, just a couple of months in 2008. This may be all the more the case, since the unsupportive physical market fundamentals all too obviously reasserted themselves very quickly after July that year. Yet it is precisely this window of illogical oil price behaviour which most clearly demonstrates how speculators took over the oil market in the first half of 2008 to push prices ever higher – regardless of the fundamentals Ruehl sees as dominating price formation. And, as we have seen, the subsequent abrupt collapse in the oil price is also in itself further evidence of a speculative bubble having been blown, this time in the characteristics of its unwinding.

I hope I have demonstrated that there is ample evidence that the well-documented explosion of institutional investor interest in a wide range of speculative financial exposures to the oil price in particular, and commodities in general, enjoys multiple channels of transmission onto the Nymex oil futures market – the market that determines the benchmark global crude price. Some participants such as hedge funds and floor traders are visible in regulatory terms as obviously "non-commercial", by definition speculative players. Yet the swap dealers are different. A class of market participant instrumental in transforming off-exchange financial interest, going long of the oil price, into matching on-exchange positions, these are the real "stealth bombers" of speculation. They are rendered invisible to the conventional radar of speculator detection by dint of their traditional regulatory classification as "commercial" players, which immediately obscures any accurate measurement of true speculative interest on the exchange. But they nevertheless undoubtedly delivered and can still deliver the highest-powered weight of speculative investment ordnance onto the target, the oil futures curve. What facts are known about the over-the-counter (OTC) commodity derivatives trading funnelled onto the regulated, global price-setting market by swap dealers, indicate very significant volumes of pricing pressure.

In making this determination of speculative pressure dominating the oil market in the first half of 2008, we are confronted with a series of studies that purport to disprove this thesis. Yet none of these studies manage to make an incontrovertible case that speculation did not play a major role in moving the market through this period. Some studies fail to do so simply because their analysis does not take account of the swap dealer loophole; others because their assessment of trading patterns is too limited in scope, does not take account of basic investment fundamentals such as rebalancing and, not least, has been undermined by dissenting opinion from a regulatory source. We have also seen how many of the pathological distinctions identified by key exponents of asset bubble theory, such as Minsky and Shiller, are clearly present in the oil futures market. (These identifying marks including various endogenous market feedback loops, the logic of "naturally-occurring Ponzi schemes", and the groupthink of "New Era" paradigms.) We have further seen how the oil price through this period was clearly reacting as much to financial investor concerns over dollar weakness and inflation as it was to potential supply disruptions; and how the pattern of oil price movement up to its peak, and the subsequent sudden collapse, matches up to another (almost) undisputed market bubble, the dotcom craze exemplified in the Nasdaq market peak in 2000. Also on the margins of possibility is market manipulation as a final contributor to the heady oil price of summer 2008, as indicated by Robert McCullough in particular.

Perhaps none of this will seem so important when, in the future, oil prices are averaged out over the "long run" of, say, the first half of the 21st century by a subsequent generation of analysts and economists. It clearly does matter in the short run, however, to millions of ordinary people worldwide, but particularly the poor in the developing world struggling, even in more ordinary times, with fuel costs eating up a disproportionately large share of their meagre income. It also matters to policy-makers and central bankers struggling to determine the appropriate interest rate-setting response to perceived inflationary pressures across the global economy, where an over-zealous toughness on the cost of borrowing can choke off activity and reduce living standards for no good reason. I have already criticised the "speculation is a red herring" stance taken by, among others, *The Economist* magazine, when $100-plus oil prices were all the rage in summer 2008. Yet I cannot help but agree with the same organ's comment in its June 6 2009 issue: 'In all the talk about the American housing market and banking misjudgments, the role of oil at $140 a barrel in sparking this recession has probably been underestimated. Last year, oil may have been sending a false signal to central banks by pushing up headline inflation when the economy was already weakening.' In other words, the fear of inflation being stoked by the prevailing high oil price prevented central banks from cutting interest rates as soon as they otherwise would have to combat the current downturn, a charge that has been widely made elsewhere as well. The cost of this forced interest rate paralysis in early 2008 to the global economy is literally incalculable, not least as the contrary scenario, where interest rates were cut earlier despite high oil prices, is an unobservable historical counterfactual; but must be presumed to be material. There is a genuine possibility that reckless speculation in oil futures severely worsened, maybe even partly precipitated the traumatic economic crisis we are still living through. Ultimately, the oil price spike of summer 2008 is an illustration of how the financialisation of commodities is coming to dominate price determination in the physical markets of those all-important goods that modern industrial society simply cannot be without. Economist L. Randall Wray noted this rising influence in his 2008 Levy Institute paper *The Commodities Market Bubble*, most of which was written before the bubble burst, in other words before there was *conclusive* proof that the price behaviour Wray considers was indeed more financial speculative excess than anything else. 'To be sure,' he wrote, 'it is very difficult to determine how much fault should be placed in the laps of money managers, because there are a number of forces coming together in a "perfect storm" to drive up commodity prices. Still, there is adequate evidence that financialisation is a big part of the problem, and there is sufficient cause for policymakers to intervene with sensible constraints and oversight to reduce the influence of managed money on those markets.' Whether or not you agree with Randall Wray's call for policy action to restrict this influence, one thing is very clear – we are all still living with the consequences of our 2008 summer of petromania.

6.2 The Aftermath

If the oil price wreaked havoc on its way up, it certainly made almost as much mess on the way back down. While ordinary citizens worldwide breathed easier with lower fuel prices, oil companies themselves were knocked for six by the abrupt collapse in price expectations. Banks that had been happy to lend money to small operators on the basis of a triple-digit oil consensus suddenly slashed the finance being made available to these same firms as they came back up for refinancing. Several went straight to the wall, among the most notable the dual Canadian/UK-listed North Sea explorer Oilexco. Oilexco's fate was poignant as it had just been responsible for probably the best quality new UK North Sea oil discovery of the decade, in the Huntingdon prospect. But Oilexco's bankers refused to extend fresh credit for the company to continue with its development plans, and as current cash from production would not suffice to refinance existing debt, Oilexco went bust. Huntingdon has now been snapped up in a fire sale of Oilexco licences by UK midcap explorer Premier Oil. Good for Premier, but tough on Oilexco and its shareholders – and arguably a disincentive for future entrepreneurial interest in wringing more out of the UK's remaining, but dwindling, oil and gas endowment.

Has such collateral damage wrought by the oil bubble in its implosion compromised to any further degree existing UK energy independence or energy security? Or may it likewise have compromised the same considerations for other countries or the global economy as a whole? Backing out to the wider picture, there seems little doubt that overall investment in oil and gas development internationally has dropped as a result of the low prices prevailing at the turn of 2009. International oil majors such as Shell or BP have slowed down project development and slashed capex plans for the next couple of years. Indeed, this is to such an extent that many analysts feel that one effect of the price blow-out has been to worsen the prospects for fresh supply emerging in the near-term. Unconventional projects, so important to most forecasts gainsaying peak oil fears with expectations of increasing supply into the 2020s, have been particularly hard hit. Shell has put off investment approval on a large expansion in its Canadian oil sands production. Some would say this is a good thing given the environmental damage oil sands production wreaks on its locale, but nevertheless slower oil sands development on the part of one of its major proponents spells a downgrading on fundamental supply expectations for the next few years.

The crash has also strengthened the grasping hand of resource nationalism. Share prices across the oil sector plunged worldwide as the oil price did. Many companies lost access to equity markets at any cost and so were stalled in development terms, marooned on promising deposits they do not have the cash to develop. Post-price crash, there is still merger and acquisition action ongoing in the sector – but it is distinguished by being driven by state sector players able, in the current environment, to pick and choose their targets, and avoid paying excessive premiums. Private sector corporate players are, after all, battening down the hatches and pulling in their horns while the ravaged banking sector remains on life support. Why should the larger private sector oil companies buy more opportunities when they are delaying funding the ones they have already? Into this vacuum step cash-rich state sector oil companies from emerging economies, who are picking off listed, mid-sized producers with decent reserves left, right and centre.

Notably, there is nary a squawk from shareholders that the offers aren't high enough. In this environment they know there is little chance of a private sector white knight riding forth with a better offer. Thus it was that while shareholders in AIM-traded SIBIR Energy, which shares production from a Russian oilfield with oil major Shell, saw their stock trade as high as 839p in mid-2008, a year later they were happy to accept a 500p per share cash-out from state-owned Russian company Gazprom Neft – frankly, an ignominious end to a previously very promising independent Russian oil producer. But a welcome outcome, one would think, for the state interests in Russia pushing for ever tighter control of "strategic" sub-surface resources. Meanwhile Addax Petroleum, another dual Canadian/UK-listed explorer, was bought in June 2009 by Chinese state oil giant Sinopec. Geopolitical analysts are still struggling to compute what the intervention of China itself in endorsing and bolstering regional Kurdish authority over the fields in northern Iraq – seen as Addax's jewel in the crown – means for the balance of power in that country between the autonomous Kurdish administration and the central government in Baghdad. This is because the central Iraqi government disputes the Kurdish right to administer its own oil licences, such as that awarded to Addax several years ago. But China has certainly been quick to tie up resources going cheaper than they otherwise would, as a result of the price crash.

When it comes to physical market fundamentals, the effect of the oil price boom-and-bust has clearly been to aggravate many of the key factors that were themselves feeding into L. Randall Wray's 'perfect storm' the first time around in any case. If the majority of people on this planet desire and indeed need a measure of stability in their energy costs, then the experience of 2008 was not a good indicator that the financialised markets that determine the oil price are working to achieve this effect. But it has not only been oil project developments, oil sector share prices, forecasts regarding future oil supply additions, and efficient market economic theories that have suffered a serious mauling as a result of the oil price bubble's explosive pop. Reputations have been shredded as well.

6.3 Goldman Sachs: Panto Villain?

If the petromania of 2008 was a stick of rock, the name running all the way through it on the inside would be that of US investment bank Goldman Sachs. As a *Financial Times* headline put it one day near the height of the fever, 'Goldman's analysts speak, and the price of crude rises'. Bullish Goldman Sachs analyst forecasts on crude prices are seen by many to have been instrumental in pushing oil futures higher at crucial points in the ramp-up to the July peak – particularly the crazy weeks for the futures curve, as detailed above, which came on the heels of their May 16 research note. But Goldman Sachs and its diverse spread of interests appear at every stage of this story, far beyond simply being the analyst cheerleaders for ever-higher oil prices.

Through its control of commodities trader J. Aron & Co., the bank is a trader and market-maker in physical crude oil – it can easily get its hands on the black stuff itself, if need be. This same trading company is a presence on Nymex, recognised as a commercial rather than a non-commercial market participant. On Nymex, however, Goldman Sachs is not just acting as a physical oil trader seeking to hedge commercial risks but also a pre-eminent swap dealer hedging out exposures from writing the sort of off-exchange, OTC derivatives that form the "dark matter" of oil pricing. It is, as research by Greenwich Associates has established, joint top bank for peddling off-exchange OTC commodity derivatives both to commercial market players and financial investors (alongside fellow Wall Street heavyweight Morgan Stanley).

Meanwhile Goldman Sachs is a leading "prime broker" for the hedge funds that are likewise so conspicuous in commodity price speculation, in essence acting as an institutional "sugar daddy" who lends such funds money in the first place, facilitates their trading and settlement, handles all of the tedious back office administrative stuff, and possibly even provides strategic research and investment advice. And, of course, Goldman Sachs did not only invent in the first place the whole concept of the commodity index when it launched the GSCI in 1991. It is also widely known, although the CFTC itself has never publicly identified it, as the bank that first importuned the regulator to grant an exemption on position limits to swap dealers hedging-out exposure incurred from enabling institutional investors to take bets on a commodity index. In other words, the "swap dealer loophole" was originally instituted by the CFTC for the benefit of Goldman Sachs.

It is therefore understandable that the heavyweight presence Goldman Sachs spreads across several sectors of the oil market means people can't help but conclude this preponderance translates into some kind of trading advantage. The *Washington Post* put it politely, in yet another piece noting how oil prices jumped on the back of that Goldman note in May 2008: 'Some say Goldman – which acts as an oil broker, runs the biggest commodity index fund, provides investment advice and trades oil on its own account – has too many institutional conflicts of interest.' Oil sector consultant Fadel Gheit put it far less politely and more directly in a June 23 2008 hearing, when US Congressman Bart Stupak, referring specifically to Goldman as the 'largest commodity dealer on Wall Street' and noting its $200 per barrel forecast, asked Gheit whether there is an 'actual or apparent conflict' in an investment bank talking up the price of commodities through its research arm, whilst speculating on it at the same time. 'Absolutely, unequivocally,' answered Gheit, 'It becomes a self-fulfilling prophecy when the largest trader predicts the price of a commodity. Guess what's going to happen to the price? It's going to follow the leader.'

Of course, joining the dots in this fashion would be a no-brainer for anyone observing Goldman's role across the oil sector. Except for the fact that any such systematic leveraging of information relating to client instructions received, or advice dispensed, across the whole group would be highly illegal. Preventing the proprietary trading arms of banks from "front-running" in markets on the basis of knowledge regarding how clients intend to trade, or profitably pre-positioning themselves in markets in expectation of house research producing a predictable reaction on the part of clients, is the whole point of the so-called "Chinese walls" that are maintained as a regulatory necessity between different departments of the same financial institution. Imputing that anything else happens is a very serious charge with regard to any financial institution, and Goldman Sachs has had to develop a thick skin in this regard. The allegations levelled at it run far beyond the sort of "nudge-nudge, wink-wink" suggestions that it may occasionally benefit from putting together information from disparate and ostensibly ring-fenced operations.

In April 2009, US business magazine *Forbes* ran a story titled 'Did Goldman Goose Oil?' The main thrust of this piece was to detail suggestions that the investment bank's traders had deliberately manipulated markets to achieve the July spike in the oil price, simply in order to bankrupt the US oil trader SemGroup. The reasoning runs, firstly, that bankers from Goldman had gained knowledge of the precarious position SemGroup had put itself in by going short of the oil price in a series of derivative trades. This Goldman discovered when they inspected its financial health prior to a planned private placement of equity in the company, which the bank was deciding whether or not to support (in the end this deal did not come about). Theoretically, according to *Forbes*, an awareness of SemGroup's vulnerability could have motivated a plan for Goldman traders to ramp up the price of oil, in expectation of therefore making the company go bust, although the piece does not actually make clear why exactly Goldman might wish to do this. It simply notes that the Goldman Sachs/J. Aron complex was both a debtor and creditor of SemGroup in different respects due to other market relationships, and implies that Goldman might expect to come away with trophies from a bankrupt SemGroup.

For its part, Goldman Sachs maintains that Chinese walls are punctiliously respected between its many departments. They aver that, in practice, this genuinely means traders elsewhere across the varied business lines in the bank do not have access to notes penned by Murti, Currie and their colleagues in equity and commodity research, before they are released to external clients. Goldman emphasises that such interdepartmental impermeability was naturally also the case between any bankers who might have gained some knowledge of the SemGroup futures exposures, in preparing the aborted deal, and its own oil futures traders operating on Nymex. Of course, one would not reasonably expect the bank to say anything else – but then this illustrates the "damned if they do, damned if they don't" situation Goldman Sachs faces with regard to the charge that it profited improperly from what went on with the oil price in summer 2008. Like the accused in a mediaeval witch trial, there is nothing it can say that will demonstrably prove its innocence as far as the inquisitors are concerned.

Nevertheless, *Forbes* seems to have got it wrong about Goldman and SemGroup. The players quoted in the piece, suggesting such malpractice might have happened, all seem themselves to be locked into an adversarial legal face-off with Goldman over the collapsed SemGroup assets. This arguably, perhaps obviously, colours their opinion of the bank. Further, an independent court examiner report, cited in the *Forbes* piece as a pending key contribution to the argument, was made public after the story was published. Far from supporting any conspiracy theory alleging Goldman manipulated the oil price to bankrupt SemGroup, it found instead that SemGroup had recklessly exposed itself by taking out short positions in oil far in excess of its own physical delivery capability, and was an accident waiting to happen if oil prices rose persistently for any reason, fundamental or not. SemGroup has only got itself to blame for going bust. Meanwhile, even analysts who disagree with him praise Jeff Currie's acuity and integrity, and acknowledge his take on oil prices as a valid one even if it not shared by themselves. Currie and Murti have been developing their view of the oil market for years, and have been consistent in their views from long before summer 2008 and the SemGroup collapse. Currie and Murti did not suddenly start ramping-up oil forecasts out of the blue.

Still, it is easy to see how, much as a pantomime villain, Goldman is painted as being "behind" everything to do with the black gold blowout. Its obvious influence across practically every crucial junction in the financial flows determining the oil price globally – from actual trading, through influential analysis, to facilitating investment for third parties on the basis of such analysis – invites presumptions of "unfair" trading advantage in these markets. And it is hard to imagine any kind of evidence that would ever be seen as sufficient to "prove" its innocence for some people. In truth, its preponderance is simply ample illustration of how closely the modern financial sector grasps oil for its own purposes.

That is not to say, however, that Goldman has not benefited from the whole oil price farrago more than other institutions might have. From a clutch of global investment banks declaring surprisingly good first quarter earnings in 2009, Goldman Sachs stood out as once more deserving its crown as the top Wall Street player. True to its "black box" reputation, it gave little detail on the specific elements contributing to such a surprise on the upside. Institutional investment sources say, however, that a large contributor to better-than-expected financial results for some top banks – Goldman particularly – was a stampede of financial investors rushing to unwind commodity positions they had previously taken out using these same banks as brokers. The gains on margins, plus break or restructuring fees realised in the course of reversing these positions for their clients, were an unexpected bonus for investment bankers deeply involved in peddling all such commodities exposure. And none were or are more so involved than Goldman.

At the same time, it hasn't all been good for Goldman. Its previously enviable reputation in calling the oil price correctly took a hammering as oil plunged. Punctuating the drop in the oil price through late 2008 were numerous media mentions of how Goldman had to completely back out of the oil price investment recommendations it had previously made to clients, leaving them high and dry in positions the bank no longer forecast any profit from. Goldman was understandably reluctant to do this. It hung on for a while, maintaining as late as October 2008 that clients should keep these trades on in the expectation that continuing "high volatility" would provide less damaging exit levels for these loss-making positions. But in late November '08, the research team led by Jeff Currie finally threw in the towel. As they wrote on November 20: 'The volatility in the past few weeks has mostly been to the downside and the pressure on the oil complex has increased. In the near term, we do not expect significant upside potential and as a consequence we are closing all of our oil trading recommendations.' The influential FT Alphaville blog run by the *Financial Times* (http://ftalphaville.ft.com) immediately ran the headline 'It's official, Goldman capitulates on oil'. Underneath, the journalist parsed the meaning of the investment bank's note: 'Translation: We were wrong and we're sorry. Ouch'.

Interviewed by me in late May 2009, however, Jeff Currie maintains that he and his team did not suffer undue blowback from irate investors. He says, 'In terms of the market collapsing, I think most people still believe in the story: if we had not been rocked by such a significant event we would have got there. After Lehman collapsed, the world shut down, there was a loss of trade credit due to the need for cash – it is hard to replicate that kind of demand shock.' And Currie also believes that once we emerge from the current economic downturn, market fundamentals will once again mandate rocketing oil prices: 'The day of reckoning is going to come – it will be a lot more painful than the financial experience and take longer. We keep lurching from a financial to an energy crisis, and because the cost of dealing with the financial crisis is so great, as soon as we come out of it we will be right back into the energy crisis.' As noted in the previous chapter, however, Currie did acknowledge that prices at their height in summer 2008 had got 'too far, too fast' in relation to fundamentals. And other commentators have been scathing of Goldman's oil price predictions post-crash. When in early June 2009 the bank upped its end-2009 price forecast from $65 per barrel to $85 and instituted an end-2010 target price of $95, a bald comment from analyst Nauman Barakat of rival investment bank Macquarie was also quoted by the press: 'Goldman's reports have lost a lot of their credibility as a result of being so wrong.'

In the end, this is probably where the most damage has been done to the bank synonymous with triple-digit oil prices. Meanwhile, the major flaws in the way the oil markets work from which Goldman can be signally seen to benefit – for example, the swap dealer loophole – are more obviously deep-seated structural issues that cut across all investment bank players, regardless of their relative infamy, rather than openings for potential criminality by particular players. Indeed, perhaps the more scandalous point is that Goldman does not *need* to engage in illegal acts to make pots of money hand-over-fist from oil price gyrations both on the way up and the way down, because it already does so simply by virtue of how these markets currently function legally. Unproveable allegations of particular misdeeds directed at Goldman

Sachs might sell well at newsstands but are a distraction from the structural market reforms required if fresh outbreaks of petromania are to be avoided. Given the position of the Nymex oil futures market at the apex of oil price formation, this would be the CFTC's job – and it now seems to be taking this job more seriously than it did in early 2008.

6.4 The CFTC Finally Sees the Light

If Goldman's reputation for forecasting has been tarnished by the whole oil price bubble, the credibility of the US futures market regulator CFTC practically evaporated in the latter half of 2008 as more and more evidence emerged underlining how much influence speculative finance actually played in determining the oil price. We have already mentioned the implication from the data presented in the *Interagency Task Force Report* issued in July 2008 (that some 80%-plus of interest across the oil futures market could be classified as potentially speculative), and how it was picked up with widespread incredulity by the media. And also how, around the same time, the CFTC was forced to reclassify one large market participant, Swiss-based oil trader Vitol, from the commercial to the non-commercial segment of the market – after an investigation found that the size and nature of the positions Vitol was taking on Nymex were more properly seen as a pattern of speculative trading, rather than the trader's previously-presumed role of facilitating transfers of physical oil in the real world. This well-publicised incident seemed to make a mockery of all the CFTC's previous protestations that it was sure it was physical fundamentals directing the price-setting process on Nymex.

Throughout the US Congressional hearings, the CFTC was excoriated by elected lawmakers for its failure either to get to grips with the influence many feel it is obvious speculation plays in Nymex oil futures trading, or to even ensure both itself and the public were equipped with the necessary level of detail on the market to have an informed argument about the issue. The "speculation is a sideshow" stance of the CFTC was also undermined more recently by one of its regulatory counterparts in the US government, the powerful Federal Energy Regulatory Commission (FERC), which monitors gas and power pricing by utility companies operating in the US. While the FERC does not officially concern itself with oil prices as such, since US gas prices are invariably linked to oil prices its views on whether or not gas prices are being driven by fundamentals also imply a view on oil market fundamentals. Gas prices in the US peaked on July 3 2008, just days before oil itself. And, of course, natural gas was also targeted as an investible commodity, alongside oil, in the commodity index trades that speculators flocked into in recent years.

So eyebrows were certainly raised in April 2009 when the FERC pointedly refused to agree with the CFTC that speculation in these commodities was not a key influence in the pricing patterns seen in 2008. In its *State of the Markets 2008* report released that month, its latest annual update on how the markets under its oversight operated through the previous year, the FERC said that: 'we believe physical fundamentals alone cannot explain natural gas prices experienced during the first half of the year… we will describe how the financial crisis that hit the country during the second half of the year altered the role of financial products and players in energy markets and increased the cost of capital while simultaneously reducing the access to capital.' In other words, the FERC thinks speculative financial investment in commodities pushed prices up, and this pressure only abated as the full impact of the credit crunch decimated the investment capacity of the financial sector itself.

The FERC concludes in its April 2009 report:

> In summary, while physical market fundamentals, particularly storage levels, can explain why natural gas prices rose during the first six months of 2008, none of the market fundamentals were extreme enough to explain why spot Henry Hub prices reached $13.31/MMBtu by July 3. As we discussed at the Winter Assessment last fall, the rise in natural gas prices coincided with a global increase in many commodity prices. This increase in commodity prices occurred as large pools of capital flowed into various financial instruments that essentially turn commodities like natural gas into investment vehicles. Ultimately, we believe that financial fundamentals along with the modest tightening in the supply and demand balance for gas during the first part of 2008, explains natural gas prices during the year.

The focus by the FERC on what it terms *financial fundamentals* in explaining commodity price spikes in 2008 is completely at odds with the oft-repeated CFTC view that physical market fundamentals alone can explain energy futures movements through this period. Many reckon this conclusion by the FERC, widely seen as the most powerful government player charged with oversight of the US energy markets, exposed the CFTC as being out on a limb in maintaining its previous stance on speculation, even among its regulatory brethren.

One year on, and it seems change is finally afoot at the CFTC. Walt Lukken, who was acting CFTC chairman through the 2008 price blow-out (and can, as a result, be personally associated with the stubborn refusal to consider speculation as a key cause of this), has moved on. He first stepped down as acting chairman in January 2009 (at which point Commissioner Michael Dunn took over as acting chairman until a new presidential appointee could be confirmed), and resigned completely as a commissioner at the agency in June 2009. This exit is in truth understandable as part of the normal run of things, as appointments to

CFTC commissioner positions are ultimately a party political issue (although subject to cross-party Congressional confirmation hearings) and the party in power in the US has switched over from the Republicans of George W. Bush, who appointed Lukken, to the new Democratic regime under President Barack Obama. The new CFTC chairman is Gary Gensler, confirmed in this office at the end of May 2009.

Speaking on June 2 before the United States Senate Subcommittee on Financial Services, only six days after taking office, Gensler made a decisive break with the previous stance taken by the agency under his predecessor. 'I believe,' he stated, 'that commodity index funds and other financial investors participated in the commodity asset bubble. Notably, though, no reliable data about the size or effect of these influential investor groups has been readily accessible to market participants. The CFTC could promote greater transparency and market integrity by providing further breakdowns of non-commercial open interests on weekly "Commitments of Traders" reports. The American public deserves a better depiction of the marketplace. The temporary relief from higher prices does not negate this need, especially given that a rebounding of the overall economy could lead to higher commodity prices.' In other words, it is official – the CFTC now recognises that the oil price movements of 2008 were indeed a bubble. Better late than never.

In various speeches since Gensler's confirmation as CFTC chairman, both he and fellow CFTC commissioners have announced a push for greater transparency and regulation of the over-the-counter derivative markets feeding into the Nymex futures curve, and also for more detailed reporting formats for the trading data regularly presented to the public. This would show more clearly the patterns of trading determining oil price formation on the Nymex. At the time of writing in mid-2009, a second report is apparently pending which will give a further update on the data collected from the special call issued to swap dealers in May 2008. Feedback has also been collected on a consultation document regarding possible changes to the "swap dealer loophole" provisions – the provisions which have allowed this class of market participants to grow to such dominance over the rest of the oil futures market.

Probably the most significant action to date, however, came just as this volume was being prepared for press. On July 7 2009, Chairman Gensler announced that the CFTC would hold its own series of public hearings from late July 2009 into August. The evidence of these hearings, and any regulatory outcome which results, will only be known once this book is printed. This is, unfortunately, always going to be rather the nature of things, with such a fast-moving story. The indications are nevertheless that some serious changes in the commodity futures trading regime are being considered.

According to the latest (July 21 2009) CFTC release on these hearings, their agenda will include garnering views from industry participants and academics on the following issues:

- Applying position limits consistently across all markets and participants, including index traders, managers of exchange-traded funds (ETFs), and issues of exchange-traded notes (ETNs);

- The effect of position limits on market function, integrity and efficiency;

- The effect of position limits on facilitating the risk management of clearinghouses;

- Whether the CFTC needs additional authority to implement such limits;

- What methodology the Commission should use to determine position limit levels for each market?

- What quantitative measures should be used in setting limits on the size of an individual trader's position?

- Should limits be established by percentage or proportion of the open interest of the market or by fixed number of allowed contracts?

- Should limits apply in all months combined, in individual months, and in the delivery month?

- How should timespread trades be incorporated in this calculation?

- Should the Commission limit the aggregate positions held by one person across different markets?

- Should exemptions from position limits be permitted for anyone other than bona fide hedgers for the conduct and management of a commercial enterprise?

- The statute states exemptions on position limits should only be granted to bona fide hedgers – what should the qualifying factors be for an entity to meet the definition of a bona fide hedger?

All of the above indicates that, at the very least, there will be some real debate over the merits or not of allowing speculative financial sector interest to be transmitted to the oil futures market through the existing "swap dealer loophole" described in this book, rather than continuing with the pretence that this whole issue does not affect oil prices. Time will tell whether or not the CFTC has both the will and the authority to face down what will undoubtedly be some vigorous lobbying from Wall Street's finest, to the effect that things should stay exactly as they are with regard to swap dealer privileges. Perhaps the argument will indeed be won by those who will undoubtedly say that, even granting episodes such as the petromania of 2008, the benefits of allowing such non-commercial investor interest to express a view on where oil prices should be in terms of price discovery and liquidity outweigh the disadvantages. Even when those disadvantages involve the most lemming-like behaviour possible by such non-commercial investment interests, and this behaviour actually takes over the market from physical fundamentals.

Personally I find it hard to believe that a regulator in the US, the land of free enterprise, will be able to take any steps that would effectively shut down outright and overnight the multi-billion dollar industry that commodity price speculation has become for the financial investment sector. And this is plainly what would happen if the swap dealer loophole was closed completely. Given that the US is also the land of litigation at the drop of a hat, presumably any such move would in any case lead to a flurry of lawsuits demanding compensation for some sort of expropriation being alleged by investment banks and their clients. What would happen, for instance, to those positions already being held under the loophole, apart from being forcibly closed and probably with many losses claimed as a result? A more likely outcome might well be that the regulator claims a discretionary power to impose particular limits on particular traders at particular times when they are seen to be breaching some sort of threshold for "systemic" influence on oil price formation, perhaps expressed as a percentage of open interest. This sounds messy and hard to work in practice, but would be in line with the more discretionary powers many advocate for banking regulators post-credit crunch; these are also predicated on ongoing monitoring of the shifting levels of systemic risk that individual banks pose at any given moment.

What does definitely, at long last and certainly not before time, seem to be happening in the interim is the replacement, in due course, of the traditional publicly-available CoT data in its current form. This is supposedly being superseded by more detailed weekly information, thus helping to establish the real breakdown between what we might describe as the "unquestionably commercial" and the "potentially speculative" interest in oil futures. This is something which, as already argued extensively, the current CoT format simply fails to capture. CFTC chairman Gensler has already announced in early July 2009 that in future CoT reports the current "commercial" categorisation will be disaggregated and a specific swap dealer categorisation detailed; and also that from the current "non-commercial" categorisation, "managed money traders", will also be disaggregated, showing hedge funds as a specific category in themselves.

To be really effective, this new standard of information would arguably have to not only monitor swap dealer activity *per se* on the Nymex (and possibly other exchanges, the ICE being the obvious candidate), but also monitor how much of the OTC business itself the swap dealer was transacting was due to commercial parties and how much due to financial investors. This does not seem to be happening thus far with the rejig of the weekly CoT data, but then again it is probably too much to hope for right now in terms of weekly organisational surveillance capability. On the other hand, Chairman Gensler also said in early July 2009 that he intends the sort of detailed data on swap dealer client exposure yielded in the special call from mid-2008 to continue to be regularly updated and aggregated by the CFTC, with a view to publication of such data in a regular quarterly update. As with the proposed changes to the CoT report, this will be a great step forward in more accurately gauging the true influence of financial investors' speculation on oil price formation in the Nymex futures market.

Regardless of what actually happens in terms of position limit exemptions, these new standards of data in themselves may help dampen, to a certain extent, excessive oil price speculation in either direction. We can imagine a "gold standard" of futures trading transparency which, if practically impossible in the near term, is certainly something that these new changes are a step towards. If market participants had sufficiently detailed information when evaluating price movements, they would be well-armed against participating in obvious and ultimately fruitless speculative bubbles. If they knew that, for example, a notional surge in the oil price could be associated with a new weight of money landing across certain maturities, due to a swap dealer hedging out exposure to financial investors – as opposed to hedging out exposure to airlines managing fuel price risk – they would be more wary of also piling in lest they turn out to be the "greater fool" that speculative bubbles rely on for momentum.

For now, however, and pending the results of the CFTC hearings set for late July and August 2009, the Nymex oil futures market lying under that agency's regulatory purview continues to operate in exactly the same fashion as it has over the previous few years. One of the fruits of this is that, incredibly, given that the last oil price bubble only burst a year ago, it seems exactly the same forces are at work again blowing a fresh bubble. But this time around, the speculatively-driven nature of this process is obvious to all but the most blinkered commentators – because through the first half of 2009 in which this has occurred, a lack of support from physical market fundamentals has been plain for all to see.

6.5 The Bubble is Burst, Long Live the Bubble

'Things fall apart; the centre cannot hold' – W. B. Yeats' famous line in his poem "The Second Coming" well describes what happened in the oil market in the second half of 2008. Prices crashed fiercely from the July 2008 all-time high of $147 per barrel. By Christmas, when a barrel of the black stuff cost just $34, previous analyst predictions of permanent triple-digit prices looked like a bad joke. The price collapse was the most spectacular ever seen in a single year since oil futures started trading on Nymex. Since then, however, the oil price has enjoyed its own "Second Coming", doubling since that trough to current levels close to $70 in late July '09, having reached this level in early June, dropped back to $60 and then climbed again since then. Once again, the oil bulls are lowing – the apparent early recovery of emerging Asian markets, supposedly evidenced in particular by China's commodity import frenzy over recent months, is seen as vindicating their "super-cycle" theories, and oil should benefit accordingly. Goldman Sachs, the überbull of oil price forecasting throughout the 2008 bubble, has as noted above once again been pushing higher and higher forecasts, notwithstanding some degree of scorn towards its pronouncements this time around.

In reality this oil price ramp-up has looked, if anything, even more out-of-touch with physical market fundamentals than the 2008 price spike. At least in the first half of 2008, no matter how much evidence there was that the emerging recession was going to send global oil demand into a tailspin, oil bulls could always tout what-if implications of sudden supply disruption meeting an obviously minimal level of headroom in the global oil supply system. Not least because, in early 2008, surplus production capacity across the OPEC oil producer cartel was seen at a mere 1m barrels per day (mbpd), a 30-year low. A year later, with global demand having tanked from perhaps 86mbpd to around 83mbpd in the first quarter of 2009, OPEC spare capacity in that same first quarter this year was around 4mbpd. That is a large cushion of safety, giving the lie to any concern over near-term physical shortages. And to experienced industry-watchers, this level of spare capacity should itself in turn be a bearish portent for oil prices in the near future.

As June 2009 rolled around, BP once again released its annual update for its *Statistical Review of World Energy*. Speaking at this event, BP chief economist Christof Ruehl said those betting on prices holding firm at prevailing (roughly $70 per barrel) levels in the near-term may have cause for concern. History shows that idle OPEC capacity does not in general stay idle for very long. Some producers in the cartel inevitably start to cheat on their quotas to gain extra revenue, with the incentive all the greater the higher prices are. Add to this the fact that the world has just seen the largest drop in oil consumption since the early 1980s; that oil stocks in storage are high in historic terms; that even despite shut-in OPEC production, current global oil supply is still exceeding demand; and the probability that in global macroeconomic terms, as Ruehl himself noted, 'a return to high rates of economic growth may prove elusive for some time'. Projections for demand growth are anaemic – in July 2009, the IEA saw oil demand rebounding from a 2009 average of 83.8mbpd to 85.2mbpd in 2010. However, from the same 2009 demand baseline, other influential forecasters are considerably more circumspect – the US government's own EIA sees 2010 demand at only 84.8mbpd, while OPEC itself is even more downbeat, with 2010 demand seen at just 84.3mbpd. Echoing the first half of 2008, it is, if anything, even harder to see how either current conditions or near-term projections for physical market fundamentals can justify the rate of appreciation the oil price has seen in the first half of 2009. Even if, as we shall examine below, the $60-70 per barrel price range might seem fair *through the cycle*, as economists say, neither current global economic conditions nor outlook have given any obvious catalyst for prices to reach the top end of this range so quickly following the trough experienced around the turn of 2008-9. Speculation, on the other hand, explains a lot.

A growing weight of speculative financial interest in the oil price has been obvious through this period, as it was leading up to oil's all-time peak last year. This time around, however, exchange-traded funds (ETFs) have been notable as a vehicle of choice for oil price speculation, as opposed to the more generalised commodity indices favoured last summer. Broker LCM Commodities pointed out in late March 2009 that one US-listed oil ETF alone, the United States Oil Fund, at that point accounted for some 15% of interest in front-month oil traded on Nymex. And UK-based global ETF provider ETF Securities said in May '09 that inflows of long interest into its suite of oil-tracking ETFs totalled $954m in the first four months of 2009, more than twice all inflows seen throughout 2008, and representing an annualised growth rate of 544%. As we know, providers of all such derivatives end up hedging their eventual net exposure onto the actual oil futures market, and hence add to pressure on the front-month Nymex light sweet crude oil price. Astonishing as it may seem that another bubble is being blown in the oil price so soon after last year's burst, the watchword has to be: 'The speculative oil bubble is burst; long live the speculative oil bubble.'

One thing at least starkly different from the 2008 bubble is a wider scepticism this time around on the part of investment bank analysts toward higher prices, and particularly prices in triple digits. This time last year, you could count on two fingers the number of top-tier investment bank oil analyst teams warning that $100-plus crude prices were cloud-cuckoo land – namely Ed Morse and his team at (now-defunct) Lehman Brothers and Colin Smith and his team at Dresdner Kleinwort. Other analysts might not have completely bought into the oil bull story, but on the other hand they felt they did not want to second-guess the market, so used the oil futures strip to value companies regardless. They thereby locked in valuations on the basis of $100-plus oil in any case. This time around, the long-time dissidents (Ed Morse and fellow ex-Lehman colleague Daniel Ahn having now relocated to LCM Commodities) are at least being joined by other big bank names as-yet sticking to a $60-$70 range for their long-term oil price in calculating company valuations. This is in defiance of whatever the futures strip might currently say to the contrary (with five-year oil having recently traded above $80 per barrel). Citigroup, for instance, is using $65 per barrel for its long-term oil price assumption, while Credit Suisse is sticking to $70. The latter bank also recently made a comment that many of its analyst peers also seem to agree with, which further crimps any justification for oil prices heading back towards the $150 they were chasing last summer. Namely, that the economic collapse experienced worldwide in 2008 has had such a serious effect that it has *permanently* impaired the likely course of oil demand growth from the levels the bank had previously seen as prevailing through the coming years.

Ironically, however, even as Credit Suisse adds its voice to those seeing greater moderation in oil prices going forward, at the same time it is feeding the very same trends that blew the bubble in the first place. The bank has also just announced the creation of its own proprietary commodity index series, to be administered in partnership with privately-owned commodity trading giant Glencore. As with any other such index, the purpose is, of course, to structure and sell investment products based on its movements. Called the Credit Suisse Commodities Benchmark (CSCB), this index too can be expected to add to the weight of non-commercial investment interest being brought to bear on the Nymex oil futures curve.

In sum, with there being as yet no firm regulatory changes enacted to ensure there will not be a repeat of the speculative excess seen through the oil price bubble of 2008, we cannot be sure that it will not happen again, and perhaps sooner rather than later. The fact that physical market fundamentals are even more clearly dissonant with the rapid price appreciation witnessed through the first half of 2009 than they were with the price action that saw crude trading above $100 per barrel a year previously, only serves to underline how easy it remains for speculative interest to get the upper hand in determining oil prices on the Nymex.

6.6 Oil Futures – a Fair Price for Oil?

What *should* the price of a barrel of oil be? Conventional economic theory tells us the answer is of course whatever the market wishes to pay. Given the global asset market events of the last two years, however, this kind of efficient market theory is looking dangerously threadbare. Most sensible people would now agree – and it seems we can now potentially include the new head of the CFTC in this camp – that speculative market bubbles can obscure the underlying fair value of an asset. Presuming we can make a distinction between the price the market is paying for a barrel of crude oil at any given point in time, and the price that could be seen as *justified* at that same time – in other words, ignoring the protests of efficient market theorists – can we say anything about where we might *sensibly* expect crude oil prices to trade around in the next few years?

BP chief executive Tony Hayward certainly thinks so. Speaking at the 2009 *Statistical Review* launch alongside his chief economist Ruehl, he repeated an argument he has used several times before to support the view that a "fair" price for oil lies somewhere in the $60-90 per barrel range. The reasoning here is that only at a floor above $60 per barrel can core producing countries in oil cartel OPEC afford to both maintain their existing oil production infrastructure at a level that sustains current output, and also meet their stated social spending commitments to their generally young, and potentially restive, domestic populations. If prices drop below this level, it is oil infrastructure spending rather than social spending which will suffer first, so then the consequent degradation of production capacity should see the oil price subsequently rise in any case back towards a level that supports both spending priorities. So much for the $60 floor, but Hayward also feels that the marginal cost of supply for emerging sources of production – such as deepwater offshore Angola and Canadian oil sands – means the upper band of sustainable prices heads towards the $90 mark.

Dr Ed Morse and Daniel Ahn of LCM Commodities have a more specific price band in mind. They have developed a model matching the five-year forward oil price to the US Bureau of Labor Statistics' producer price index (PPI) for oil and gas industry services, the five-year price being conventionally accepted as the market's best expectation of a sustainable long-term oil price.

Between January 1994 and October 2007, movement in the five-year forward WTI oil price as traded on Nymex was very closely tied to movement in the PPI index they have developed and back-tested through the historical data (in statistical terms, yielding an "R-squared" measure of correlation around 96% through this period, although Morse and Ahn caution against reading too much into such a pronouncement). Their chart as reproduced in Figure 14 details the recent history of this relationship, from January 2004 onwards, and shows the five-year forward oil price plotted against both the "plain vanilla" PPI index and also the PPI index adjusted for dollar fluctuations, to strip out the effect of exchange rates feeding into oil service supplier pricing. As can be seen, the fit between actual five-year forward oil prices and what these twinned PPI indicators said they should have been was incredibly close up until late 2007. Simply observing the visual fit, however, it is clear that in contrast to the preceding period, the actual real five-year price seriously diverged upwards from the PPI indicators after October 2007, eventually morphing into the oil price bubble which peaked in July 2008.

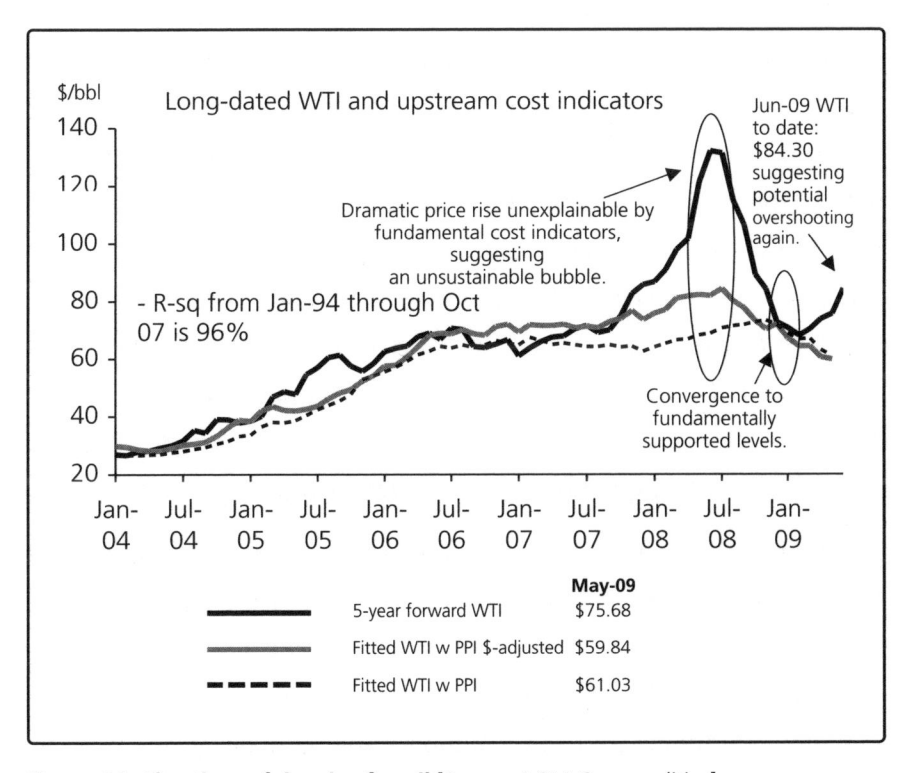

Figure 14: Charting a fair price for oil [Source: LCM Commodities]

While the five-year oil price and PPI data series briefly converged again around January this year, oil has since run away upwards again into what looks, from this chart, like the beginning of just another bubble – which fits perfectly with what we are seeing in the markets as above. Right now the PPI series predicts the "fair" five-year price in a range of around $60 a barrel.

Both Hayward and the LCM team under Ed Morse offer different reasons for pegging oil significantly below $100 per barrel for the foreseeable future. Ultimately, though, each argument comes back to concentrating on producer cost structures and the necessary returns that must be distributed through the industry to motivate development and maintenance of oil production. Such thinking is in stark contrast with the price-spike-to-infinity models that characterised the oil price fever of 2008. These models were predicated on fundamental supply deficits not capable of remedy simply by gradual price increases to bring more marginal sources into production. Instead, they were requiring prices to leap high enough to permanently destroy demand in relation to a supply which *could not* be significantly increased. As long as 4mbpd of spare OPEC capacity remains idling, however, such jeremiads will appear baseless – and current real world conditions do not indicate this buffer will be evaporating quickly any time in the near future.

I personally have my own, more cynical reasoning for thinking that somewhere in the $60-70 per barrel range is a reasonable expectation for the price of oil. As with Hayward's theory, mine is linked to the self-interest of the OPEC cartel which, in this new world of more straitened economic conditions, and a hefty supply buffer unlikely to be eroded by oil demand growth in the near-term, will once again have a lot of discretion in matching supply to a particular targeted price band. Unlike Hayward's theory, it is not as complicated as calculating the price band sufficient for OPEC members to realise both social and oil infrastructure spending. Instead, it simply notes the fact that around $60-$70 per barrel looks to be the highest OPEC can get away with managing prices to before they become high enough to incentivise fresh investments in unconventional sources of supply such as oil sands and gas-to-liquids. (These being the sources which threaten, in the longer term, to diminish the power the cartel holds over the global oil market.) Potential Canadian oil sands resources are often touted as equalling or exceeding "another Saudi Arabia". We can be sure the genuine article does not wish these resources to become economically viable any time soon. This, as much as social spending commitments, must surely lie behind frequent indications from various OPEC mouthpieces that it sees around $70 per barrel as the right price for oil.

What will the oil futures markets, however, think is the right price for a barrel of crude oil through the coming years? At the time of writing, in July 2009, we are just past the first anniversary of oil's as-yet all-time high. It is impossible to say if, by the time this book is published, real world fundamentals will have reasserted themselves to choke off the fantastic and many would say unjustified rate of appreciation already seen in the oil price year-to-date, or if indeed the price will have charged even higher on the back of the same sort of speculative frenzy that brought us that 2008 blow-out in the first place. But if the lesson of the oil price bubble of 2008 is that speculative fantasy can all too often displace the real world in determining the price of our most important commodity, on the evidence of the first six months of 2009 it is clear the oil market itself is yet to fully awake from the fevered dreams of petromania. Perhaps in a world becoming ever more financialised, ever more colonised by the imperatives of financial investment, with profit-hungry investors and institutions riding roughshod over the actual instrumental uses the real world has for the goods we trade between ourselves, complete escape from this malaise is impossible. Like a parasitic infestation lying latent in the lifeblood of the oil market, recurrent bouts will return to disorientate us. Petromania is a condition we might simply have to learn to live with.

Sources & Bibliography

Core Sources, With Explanatory Notes

For a selection of core, what we might call evidential source documentation, referred to repeatedly and in detail through the text, I have adopted a non-standard format of bibliographic listing as below. I trust this will make it easier for readers to both grasp the respective significance of each, and locate copies of the material for themselves if so desired. All other material is presented in the more standard bibliography that follows this.

At the time of writing, some of the source documents have become harder to locate than they were at earlier points in time. The CFTC website (www.cftc.gov) in particular is notably poor at returning search results linking to CFTC documents that the author himself previously downloaded from official CFTC web pages now no longer maintained in the same form. Nevertheless, where official channels fail, the internet will often deliver the same documents through either generalist search websites such as Google or, with more specific regard to academic papers, through specialist sites such as the Social Science Research Network (SSRN – www.ssrn.com).

Additionally, in the interests of both space and sanity, I have eschewed direct references here to the veritable legion of news reports from media around the world regarding the oil price and associated issues from the mid-noughties, up to the time of manuscript submission in mid-2009. A very few important and/or exemplary stories from specific media have already been dated or sourced in the text itself, and other references to news stories may be easily verified *de novo* through the use either of a generalist search website or a subscription-based news tool (such as Dow Jones Factiva).

CFTC-Originated Studies, Reports & Miscellanea

(See www.cftc.gov, with the proviso above.)

Fundamentals, Trader Activity and Derivative Pricing
(Bahattin Büyüksahin, Michael S. Haigh, Jeffrey H. Harris, James A. Overdahl, Michel A. Robe, 2008)

The latest version of a CFTC study on co-integration between nearby and long-dated oil futures prices as traded on Nymex, incidentally also detailing the overwhelming dominance of swap dealers in the contemporary market, and the source for the data detailing the split between trader categories in 2000, 2004 and 2008 used in this book. Data herein is far more detailed than that normally released to the public in the weekly CFTC Commitment of Traders (CoT) report.

Market Growth, Trader Participation and Pricing in Energy Futures Markets
(Michael S. Haigh, Jeffrey H. Harris, James A. Overdahl, Michel A. Robe, 2007)

Earlier version of the study on cointegration through the oil futures curve (as above) – trader data snapshots herein cover just 2000 and 2006.

Price Volatility, Liquidity Provision and the Role of Managed Money Traders in Energy Futures Markets
(Michael S. Haigh, Jana Hranaiova and Jim Overdahl, 2005)

The "original" CFTC study (itself a November 2005 third draft of a working paper first issued in April 2005) on whether managed money traders (MMTs), more commonly known as hedge funds, could be seen to lead price formation on the Nymex. Despite its very limited scope, this work (or a later update of it) was repeatedly referred to by then-CFTC chair Walt Lukken and other officials as "proof" that speculation played no notable role in driving oil prices through early 2008.

Herding Amongst Hedge Funds in Futures Markets
(Michael S. Haigh, Naomi E. Boyd and Bahattin Büyüksahin, undated itself but externally dated by SSRN to 2007)

Examining data running from 2002-2006, this paper is similar to that above insofar as it examines the behaviour of MMTs/hedge funds in commodity futures trading. This time it uses data from 32 separate commodity futures markets, as opposed to oil futures in particular – and concludes that while such investors do indeed exhibit "herding" in their trades, this nevertheless does not translate into speculation dominating price formation. Relevant as a later restatement of the same conclusion made in the 2005 MMT study, and therefore also possibly the research CFTC officials such as Walt Lukken were referring to when they claimed that the CFTC had proved that speculation was not an issue in oil price formation (see below).

Written Testimony of Acting Chairman Walter Lukken Before the Subcommittee on Oversight and Investigations Committee on Energy and Commerce, United States House of Representatives
December 12 2007 (CFTC, 2007)

In which Acting Chairman Lukken details to the subcommittee the findings of research 'recently' (sic) undertaken by the CFTC, the conclusions of which, as he states, 'show that speculative buying, as a whole, does not appear to drive prices up'. The research in question is not specified beyond this: it could be either of the two studies listed directly above, whether originals as named or updates, a combination of both, or another study entirely.

Letter from CFTC Acting Chairman Walt Lukken to The Honorable Jeff Bingaman, Chairman, Committee on Energy and Natural Resources, United States Senate
June 11 2008 (CFTC, 2008)

In which CFTC Acting Chairman Walt Lukken details the purpose both of the then-recently issued 'Special Call' to swap dealers trading oil futures on Nymex and the creation of the Interagency Task Force, with regard to examining the role of speculation and index trading in crude oil price formation on Nymex.

Interim Report on Crude Oil
(Interagency Task Force on Commodity Markets, 2008)

The first and, despite being 'interim', thus far only report on crude oil futures pricing issued by the special interagency task force put together and chaired by the CFTC in response to widespread criticism of its stance on speculation. Its findings that speculation played no significant role in oil price formation are not only questionable in their own terms but were also publicly disowned by CFTC commissioner Bart Chilton (see below).

Staff Report on Commodity Swap Dealers & Index Traders with Commission Recommendations
(CFTC, 2008)

The CFTC staff report on swap dealers, the first fruits (thus far) of the 'Special Call' the CFTC issued to swap dealers in mid-2008. As with the *Interim Report* (above), its findings can be questioned in their own terms. Also included as an appendix in this report is the 'Commissioner

Bart Chilton Dissent', in which the eponymous CFTC staffer disowns not just the *Staff Report* findings but also those of the *Interim Report* as well.

Testimony by Gary Gensler, Chairman, on behalf of the Commodity Futures Trading Commission Before the United States Senate Subcommittee on Financial Services and General Government, Committee on Appropriations

June 2 2009 (CFTC, 2009)

In which newly-appointed CFTC chairman Gensler reverses the official position of the oil futures trading regulator by stating that there was indeed a bubble in commodity prices and that commodity index funds and financial investors were obvious players in this.

Statement by Chairman Gary Gensler on Speculative Position Limits and Enhanced Transparency Initiatives

July 7 2009 (CFTC, 2009)

In which CFTC chairman Gary Gensler announced a new wave of (currently ongoing, at the time of writing) hearings on restricting speculative pressure on commodity price formation in futures markets.

CFTC to Hold Three Open Hearings to Discuss Energy Position Limits and Hedge Exemptions, First Hearing Scheduled for July 28 2009

(CFTC press release dated July 21 2009)

Official confirmation that the hearings announced by Chairman Gensler are underway, detailing the scope of the enquiries.

The weekly **Commitment of Traders** (CoT) report and archives thereof are available from:

www.cftc.gov/marketreports/commitmentsoftraders/index.htm

Other US Government Agency Reports

Issues Involving the Use of the Futures Markets to Invest in Commodity Indexes [sic]
(US Government Accountability Office, January 2009)

A briefing prepared by the GAO for the House Committee on Agriculture in December 2008 and publicly released in January 2009 as a letter dated January 30 2009 from Orice M. Williams, Director, Financial Markets and Community Investment, to The Honorable Collin Peterson, Chairman, Committee on Agriculture, House of Representatives, summarising the GAO conclusions alongside the briefing slides. The letter itself details the GAO view on the limitations of various studies regarding speculation in oil futures which rely only on standard CFTC CoT data. The whole document is publicly available at www.gao.gov.

State of the Markets Report 2008, Item No. A-3
April 16 2009 (Federal Energy Regulatory Commission, 2009)

The FERC review of the wild energy price volatility seen in mid-2008 maintained that physical fundamentals alone could not explain this phenomenon and focused instead on so-called "financial fundamentals", in a fashion completely at odds with the then-prevailing CFTC view that financial speculation played no important role in these events. Available from www.ferc.gov.

Submissions to US Legislative Hearings and Associated Documentation

Prepared Testimony of Philip K. Verleger, Jr.

(PKVerleger LLC 15 W. Francis Street, Aspen, Colorado 81611)

To the Permanent Subcommittee on Investigation of the US Senate Committee on Homeland Security and Governmental Affairs, and the Subcommittee on Energy of the US Senate Committee on Energy and Natural Resources

December 11 2007

Philip Verleger's late 2007 testimony recounted his evidence and analysis that the US Department of Energy policy of filling the Strategic Petroleum Reserve (SPR) with light sweet crude was squeezing WTI oil futures prices as traded on Nymex and would lead to an oil price of $120 per barrel in due course. This testimony plus much other relevant material, including Verleger's Comments on the *Accidental Hunt Brothers* report in which he dismisses Michael Masters' efforts as 'junk analysis', can be found at the PK Verleger LLC website (www.pkverlegerllc.com).

Testimony of Guy Caruso

Administrator, US Energy Information Administration

To the Subcommittee on Energy of the Committee on Energy and Natural Resources, and the Permanent Subcommittee on Investigations, of the Committee on Homeland Security and Governmental Affairs, United States Senate

December 11 2007 (US Department of Energy, Office of Congressional & Intergovernmental Affairs, 2007)

Written testimony notable for the manner in which Caruso defers to CFTC judgment on whether or not speculation is a major factor in oil price formation (explicitly quoting the CFTC work on MMTs/hedge funds), a stance which oral questioning from senators confirmed Caruso maintained despite his own feeling that speculation was playing at least some part in high oil prices (see below).

Senators Carl Levin and Byron L. Dorgan
Joint Hearing on Speculation in the Crude Oil Market
Committee Hearing
11 December 2007 (CQ Transcripts, 2007)

Transcript of same US Senate hearing referenced above. Contains in a single record of one legislative hearing: comment from oil consultant Fadel Gheit of Oppenheimer & Co, Inc, that then-current oil prices were being inflated by speculators to levels twice what they otherwise might be; comment as noted above from oil consultant Philip Verleger of PKVerleger LLC that the Strategic Petroleum Reserve (SPR) policy would drive crude prices to $120 per barrel; and a long grilling of EIA administrator Guy Caruso by various senators in which they force him to admit that while he himself thinks speculation is indeed a cause contributing to high oil prices, he nevertheless defers to the CFTC judgment that it should not be seen as a significant contributory factor.

Testimony of Michael W. Masters
Managing Member/Portfolio Manager, Masters Capital Management, LLC
To the Committee on Homeland Security and Governmental Affairs, United States Senate
May 20 2008

An appendix included in the submission from Masters to this hearing was the first time he laid out before legislators his methodology and resultant calculations regarding the volume of long, buy-side interest that commodity index investment was bringing to the oil futures market.

Testimony of Roger Diwan

(Partner at PFC Energy, 1300 Connecticut Av, NW Washington DC 20010)

Before the United States House of Representatives Subcommittee of Oversight and Investigation, Committee on Energy and Commerce, regarding *Energy Speculation: Is Greater Regulation Necessary To Stop Price Manipulation? Part II*

June 23 2008 11:00 a.m

Roger Diwan's written submission to the June 23 hearing includes his account of what he described as the three stages of oil market financialisation, his labelling of oil as the 'new gold' due to its emergent price correlation with dollar weakness, and his call for greater regulation and transparency in the futures market itself.

Testimony of Fadel Gheit

Managing Director & Senior Oil Analyst, Oppenheimer & Co. Inc.

To the Subcommittee on Oversight and Investigations, Committee on Energy and Commerce

June 23 2008

As quoted extensively in the text of this book, Gheit's submission lays out a clear case that when we have eliminated all fundamental reasons for the explosive price appreciation seen in crude oil futures from late 2007, speculation is the obvious answer.

Financial Energy Markets and the Bubble in Energy Prices: Does the Increase in Energy Trading by Index and Hedge Funds Affect Energy Prices?

Executive Summary, Testimony Before the Subcommittee on Oversight and Investigations of the Committee on Energy and Commerce, US House of Representatives

By Edward N. Krapels, Special Advisor, Financial Energy Markets, Energy Security Analysis, Inc., Wakefield, Massachusetts

June 23 2008

Edward Krapels essentially agreed with his peers Diwan and Gheit that it was speculation that was responsible for oil prices above $100 per barrel.

Testimony of Michael W. Masters

Managing Member/Portfolio Manager, Masters Capital Management, LLC

To the Committee on Energy and Commerce Subcommittee on Oversight and Investigations, United States House of Representatives

June 23 2008

Excerpts from the submission by Michael Masters to the June 23 hearings have been quoted earlier in this book – this evidence also reiterated his calculations, previously laid out in May 2008, of the volume of commodity index-originated long interest in oil futures.

Special Note

Testimony from the four witnesses listed immediately above – Diwan, Gheit, Krapels and Masters – formed Panel 1 of the US Congressional hearing before the House of Representatives Committee on Energy and Commerce's Subcommittee on Oversight & Investigations held on June 23 2008. Other testimony across three other panels that day included submissions from US industry representatives, including airline executive Doug Steenland as quoted earlier in this book, and testimony from market and regulatory officials. Copies of submissions from all of these witnesses plus supplementary material are available from the website archives of the US House of Representatives Committee on Energy and Commerce, in the 110th Hearings section. The supplementary material also includes an archived multimedia recording of the June 23 2008 hearing, from which are taken the various verbatim quotes from actual oral testimony related earlier in this book. The hearing in question is entitled "Energy Speculation: Is Greater Regulation Necessary to Stop Price Manipulation? – Part II" and the title/summary page linking to all these assets is currently accessible at: http://is.gd/2rBD9.

Depending on 19th Century Regulatory Institutions to Handle 21st Century Markets

United States Senate Energy and Natural Resources Committee (McCullough Research)

September 16, 2008

Testimony from Robert McCullough, Jr., of McCullough Research, quoted earlier in this book and detailing his assessment of similarities between the price patterns seen in the Nymex oil market in June/July 2008 and the price patterns seen in US West Coast power markets some years earlier, the latter now known to be caused by illegal market manipulation on the part of now-defunct energy company Enron. A wealth of other useful material from McCullough, including the more detailed report on the same subject entitled *Seeking the Causes of the July 3 2008 Spike in World Oil Prices*, originally released in August 2008 under the aegis of US Senators Maria Cantwell and Ron Wyden, is available from the McCullough Research website at www.mresearch.com.

The Accidental Hunt Brothers / The Accidental Hunt Brothers Act 2

Two reports on commodity index speculation and the resulting effect on futures prices, written by Michael Masters of Masters Capital Management – included in this sources section by dint of the fact that the original report, at least, was released in September 2008 under the aegis of US legislators who had been prominent in holding hearings on the issues, specifically Senator Byron Dorgan, Senator Maria Cantwell and Representative Bart Stupak. The title references the Hunt brothers who notoriously (and deliberately, as opposed to accidentally) cornered the silver market to cause a price spike in the early 1980s. Both reports are just some of the material available at a website dedicated to the issues surrounding commodity index speculation and maintained by Masters and associates at accidentalhuntbrothers.com.

Selected Relevant Investment Bank Analyst Reports

(Note: investment bank analyst reports are in general restricted to clients, with some additional distribution to selected media.)

A lesson from long-dated oil: A steadily rising price forecast
Jeffrey Currie et al, Goldman Sachs Energy Watch
May 16 2008

The much-referenced Goldman note from mid-May 2008 which introduced the second half 2008 average price forecast of $141 per barrel and was itself seen as kicking off a couple of hectic weeks of speculative activity in the oil futures markets.

"Is it a bubble?"

Edward Morse et al

Lehman Brothers Energy Special Report

May 16 2008

Released the same day as the notorious Goldman note above, Dr Ed Morse and team warn that the oil price does indeed look like a speculative bubble and also introduce their calculations of the leverage the oil price displays to commodity index inflows, as quoted earlier in this book.

"Oil dot-com"

Edward Morse et al

Lehman Brothers Energy Special Report

May 29 2008

The Lehman Brothers note from Dr Ed Morse and his team which laid out in detail the speculative forces causing the abnormal oil futures curve flexing seen in the latter half of May 2008, warned that oil price bulls were mistaking temporary Chinese hoarding of diesel for a long-term structural demand deficit and predicted an abrupt end to the oil price bubble.

Index Inflows and Commodity Price Behaviour

Daniel Ahn

Lehman Brothers Commodities Special Report

July 31 2008

Released as the oil price was already falling rapidly from its peak, this note reiterated in detail much of the Lehman work relating to commodity index inflows and futures pricing and is also a good, brief primer on the whole issue of financialisation in commodity markets.

South by Southeast: Recent F&D cost deflation imply continued deferred oil price weakness

Daniel P. Ahn, Edward L. Morse and Edward Kott

LCM Research Special Report

March 19 2009

This March 2009 note from ex-Lehmanites Ed Morse and Daniel Ahn in their new berth at LCM Commodities lays out both the theory and calculations of the producer price index/long-dated oil price correlation referred to in this book. It also fingers oil index ETFs as driving oil price appreciation through early 2009, and presents evidence of the commodity index outflows from June 2008 onwards which support the theory of investor "rebalancing" beginning before the oil price even peaked.

As the financial crisis eases, an energy shortage lies ahead

Jeffrey Currie et al

Goldman Sachs Energy Watch

June 3 2009

A mid-2009 note from Jeff Currie and his team foreseeing the return of a 'structural bull market' in oil pricing and instituting an end-2010 WTI price forecast of $95 per barrel.

Statistical Information on Energy Markets

BP Statistical Review of World Energy (BP)

This indispensable review is released annually by UK-listed multinational oil company BP, the current edition (2009, 58th edition) is available from www.bp.com.

IEA (International Energy Agency)

The energy agency of the OECD (Organisation for Economic Co-operation and Development, essentially the club of developed world market democracies) maintains a suite of various statistics as well as publishing its monthly *Oil Market Report* (*OMR*) and, more recently, its *Medium Term Oil Market Report* (*MTOMR*) - all of which are available via www.iea.org.

EIA (Energy Information Administration)

The EIA is part of the US government Department of Energy (DoE), and issues its own widely followed statistics on energy usage, see www.eia.doe.gov.

OPEC (Organization of the Petroleum Exporting Countries)

The oil producer cartel maintains its own statistics on oil markets and usage, all of which are available from www.opec.org.

Bibliography

Arranged Thematically

Oil, History and Geopolitics

Clarke, D., *Empires of Oil – Corporate Oil in Barbarian Worlds* (London: Profile Books, 2007)

Industry insider Duncan Clarke explains how the Western, private sector oil companies that have historically dominated the industry since its inception are facing an existential threat. The author shows how power over resources is leaching from them to rising state sector companies from outside the club of developed world market democracies.

Duffield, J. S., *Over a Barrel, The Costs of US Foreign Oil Dependence* (Stanford: Stanford University Press, 2008)

Eminent US political scientist John S. Duffield spells out the reality and implications of US dependence on crude oil from overseas.

Engdahl, W., *A Century of War: Anglo-American Oil Politics and the New World Order* (London/Ann Arbor: Pluto Press, 2004 [1992])

It is an article of faith among prominent leftist academics such as Peter Gowan and David Harvey that the hike in oil prices caused by the Arab "oil shock" of 1973 was at the least partially-inspired, at worst covertly engineered, by the Nixon administration in the US. This was supposedly both as a means of visiting recession upon oil import-dependent Europe and Japan, then-rising economic competitors to the US, and also seeing both the US dollar itself and the US bank-led dollar loan market strengthened immeasurably by a massive jump in dollar-denominated Arab oil revenues, or "petrodollars" (in reality, all these outcomes occurred). While you will search in vain for any such tale in Daniel Yergin (see below), this volume by William Engdahl is probably the best-evidenced account of this alternative history, and is endorsed by the high-profile Saudi oil minister of the 1970s, Ahmed Zaki Yamani.

Klare, M., *Rising Powers, Shrinking Planet – How Scarce Energy is Creating a New World Order* (Oxford: Oneworld, 2004 [1992])

Klare, M., *Blood and Oil* (London: Hamish Hamilton, 2004)

Klare, M., *Resource Wars, The New Landscape of Global Conflict* (New York: Metropolitan Books, 2001)

Michael Klare is arguably the preeminent academic analyst of resource nationalism and the conflict it entails. All three of these books are both extensively and comprehensively referenced enough for any professorial discussion yet nevertheless highly readable for those not steeped in the professional discourse of international relations.

Rowell, A., Marriott, M., & Stockman, L., *The Next Gulf - London, Washington and Oil Conflict in Nigeria* (London: Constable, 2005)

An early account, yet still one of few detailed examinations, of the background to the militant insurgency in the Niger Delta region of Nigeria – although violent events on the ground have snowballed considerably in the few years since it was published.

Yergin, D., *The Prize – The Epic Quest for Oil, Money & Power* (New York/London: Touchstone/Simon & Schuster, 1992)

The classic doorstopper on the history of the oil industry from the mid-1850s up to the late 20th century, replete with all the big personalities, politics, brinkmanship and war along the way, by Daniel Yergin, president of influential oil consultancy Cambridge Energy Research Associates (CERA).

Peak Oil Theorists

Deffeyes, K. S., *Hubbert's Peak – The Impending World Oil Shortage* (Princeton/Oxford: Princeton University Press, 2001)

Kenneth Deffeyes is a prominent peak oil advocate who was also a personal friend of the theory's originator, Shell geologist M. King Hubbert. This is a great introduction to the subject, in which Deffeyes also predicts 2009 as the year world oil production will hit its geologically-producible peak.

Bentley R. W., Mannan S. A., Wheeler, S. J., "Assessing the date of the global oil peak: The need to use 2P reserves", *Energy Policy* 35 (2007), pp.6364-6382

Based at the University of Reading department of cybernetics, Dr Roger Bentley is an agreeably unfanatic proponent of peak oil who will happily concede that unconventional supply sources are a wild card in any such calculations, but is nevertheless adamant that for conventional oil supply, the peak production point is imminent. This paper argues that a conventional, corporate "Big Oil"-type rebuttal of peak oil based on historic growth in proven (so-called 1P) reserves over the past thirty years is erroneous, as the focus instead should be on proven and probable (2P) resources estimated at the time of discovery for a particular oil deposit, which actually display far less growth through this period.

Bentley, R., Boyle, G., "Global oil production: forecasts and methodologies", *Environment and Planning B: Planning and Design* 35(4) (2008), pp.609-626

Another paper co-authored by Roger Bentley, this one an admirably comprehensive overview of the various schools and resulting estimates surrounding this most tendentious subject – of great use to those who both agree and disagree with peak oil theories.

Simmons, M. R., *Twilight in the Desert – The Coming Saudi Oil Shock and the World Economy* (Hoboken: John Wiley & Sons, 2005)

The controversial book by energy investment expert Matthew Simmons, questioning the production capacity of OPEC mainstay Saudi Arabia.

The Financialisation of Oil and Commodity Markets

Campbell, Patrick; Orskaug, Bjorn-Erik; and Williams, Richard, "The forward market for oil", *Bank of England Quarterly Bulletin* Spring 2006, pp.66-74

A useful survey by Bank of England staffers of both on-exchange and off-exchange, over-the-counter (OTC) oil derivative trading activity, which incidentally notes: 'In common with other OTC commodity markets there is only limited information available about the size of the OTC market for oil. But the range of market participants we surveyed were unanimous in reporting that the OTC oil derivatives market is significantly larger than the exchange-traded oil futures market.'

Domanski, Dietrich, and Heath, Alexandra, "Financial investors and commodity markets", *BIS Quarterly Review* March 2007, pp.53-67

An article issued under the aegis of the Bank of International Settlements (BIS), effectively the global banking regulator, which not only includes the estimates of financial investment in oil futures versus physical trade cited earlier in this book but is also itself a short introduction to the financialisation of commodities.

Greenwich Associates, *The Global Commodity Boom: Companies Turn to Top Derivatives Dealers for Help in Hedging*, and *Financial Investors Fueling Commodities Boom* (2008)

A twinned pair of short reports released in May 2008 by investment consultants Greenwich Associates detailing the results of surveys the firm undertook. As referred to earlier in the text of this book, these both established Goldman Sachs and Morgan Stanley as joint top dogs for energy-related OTC derivatives trading and also detailed the varying exposures of different classes of institutional investors to OTC commodities derivatives.

Hubard, W., *The Financialisation of Commodities* (London: Thomson Reuters) (2008)

This "IFR Market Intelligence" report, authored by investment luminary and sometime TV market pundit William Hubard, covers in detail the theory, practice and recent history of the commodities investment boom.

Tham, Eric, *Time Varying Factors Behind the Oil Price* (2008)

An academic paper establishing an increasing sensitivity of the WTI oil price to financial speculation since 2004 – the index of speculation in question is the net long position of "non-commercial" market participants as defined for the weekly CFTC CoT report.

Wray, L. Randall, "The Commodities Market Bubble – Money Manager Capitalism and the Financialization of Commodities", *The Levy Economics Institute of Bard College Public Policy Brief*, No. 96, 2008 (Annandale-on-Hudson: The Levy Economics Institute)

Written just prior to the great commodity price crash seen in the latter half of 2008, this paper critiques the then-prevailing commodity price boom as a speculative bubble, a natural outgrowth of the "money manager" style of capitalism inherently prone to cyclical booms and busts. Wray is a former associate of the late Hyman Minsky, working at the Levy Institute which keeps the Minskyan flame burning, who also helped prepare the latest edition of Minsky's *Stabilizing an Unstable Economy* (see below).

Speculative Bubbles and Market Irrationality

Cooper, George, *The Origins of Financial Crises* (Petersfield: Harriman House, 2008)

As referred to in *Petromania*, an excellent synthesis of various strands of financial instability theory, including Minsky's, with the author's own extensive understanding.

Mackay, Charles, *Extraordinary Popular Delusions and the Madness of Crowds* (Ware: Wordsworth Editions, 1995 [1841])

Originally published in the early 19th century, Mackay's book includes blow-by-blow accounts of some genuine investment market blow-outs – the Dutch tulipomania, the Mississippi Scheme and the South Sea Bubble – amongst a roster of more obscure episodes.

Minsky, Hyman P., *Stabilizing An Unstable Economy* (New York: McGraw Hill, 2008 [1986])

As cited extensively in this book, a core work from the man who literally wrote the book on the circular, self-reinforcing logic of speculative finance and who is now being so publicly rediscovered in the wake of the global credit crunch.

Shiller, Robert J., *Irrational Exuberance* (2nd ed) (Princeton: Princeton University Press, 2005)

Also extensively cited in this book, Robert Shiller's seminal study of market irrationality as evidenced through the 20th and early 21st century. Whereas Minsky generally concentrates on the in-built structural features of markets which push them into speculative excess, Shiller generally focuses on the behaviour of market participants in both booms *and* busts. Both aspects are required for a rounded picture of how speculative bubbles are inflated and burst.

Teitelman, Robert, "The Journal goes bubble happy", *TheDeal.com* (published May 20 2008), available at:
www.thedeal.com/dealscape/2008/05/the_journal_goes_bubble_happy.php

Robert Teitelman's pithy pathology of speculative bubbles, as quoted in this book, was published on the website of well-known US financial magazine/website *TheDeal*.

Index

A

Accidental Hunt Brothers, The 126, 204 (see also: Masters, Michael W.)

Addax 226
Ahn, Daniel 184, 246, 248 (see also: Morse, Dr Edward)
Alfonso, Juan Pablo Perez (see: 'crude, devil's excrement, the')
Al-Naimi, Ali 46, 80
Alphaville blog (see: *Financial Times, The*, Alphaville blog)
Angolan deepwater offshore 248
Arens, Richard 17-20, 101
Asia (see also: BTC; China; "commodity super-cycle"; diesel; India; resource nationalism)

    "axis of mercantilism" 38
    demand growth 21-3, 32-3, 143-5
    Emerging Asia 32-3
    oil acquisition 37-8

B

Baku-Tbilisi-Ceyhan (BTC) pipeline 31-2, 168
Bank for International Settlements (BIS) 75, 108-9
Barclays Capital 45
Barton, Congressman Joe 83
Beijing Olympics 160
Bernanke, Ben 166-7
Bingaman, Jeff 129
Blanch, Francisco 158, 216-8
Bodman, Samuel (US Energy Secretary) 134
BP 23, 39, 81-2, 224 (see also: Hayward, Tony; Ruehl, Christof)
    Statistical Review of World Energy 22, 77, 244, 248
Brent (see: crude, prime markers)

bubbles
   definition 183
   feedback loops (see: feedback loops)
   identification 180-3
   negative bubble, theory 169, 185-7
   origins 175, 179
   past bubbles 162
   speculative bubbles 170-3 (see also: speculation, bubble)
Bush, George W. 47, 237

C

Canadian oil sands 82, 225, 248
Caruso, Guy (see: EIA, Caruso, Guy)
Caspian oil 27
   attacks 31-2
Central Asia (see: BTC)
CFTC (Commodity Futures Trading Commission) 89, 91-2
   Commitment of Traders (CoT) report 96, 192, 237, 241 (see also: speculation, commercial/non-commercial split)
   Commitment of Traders classifications 124
   credibility 234-8
   Fundamentals, Trader Activity and Derivative Pricing 193
   future, the 238-9, 241-2
   Interagency Task Force 161, 193-5, 202
      Interim Report 161, 192-6, 200, 202
   Lukken, Walt (see: Lukken, Walt)
   monitoring 96
   Special Call 129-30
   Staff Report 192, 196-201, 208
   State of the Markets 2008 235
   studies, assorted 100-2, 122, 125, 149
   supervision 106-7, 119, 121
CGES (Centre for Global Energy Studies) (see: Drollas, Leo)
Chilton, Bart dissent 201-3
China 21-3, 37-8, 143-4 (see also: Asia; Beijing Olympics; diesel)

        demand growth 159-60
        hoarding 159-60, 189-90
        Sichuan earthquake 159-60
Citi 106, 246
CITIC 38
CNOOC 38
CNPC 38
commodity index investment 108, 127-9, 157, 197-8 (see also: speculation, commodities; CSCB)
        investors 150-3
commodities (see also: CFTC; derivatives)
        investment 106-8, 120
                indices (see: commodity index investment)
        super-cycle 32-3, 145, 183, 243 (see also: Emerging Asia)
swaps 117-119, 129
Congressional hearings 44, 76, 89-94, 143-5, 204-5, 208, 234-5
ConocoPhilips 82
consequences 224-6 (see also: credit crunch)
Cooper, George 179-80, 183
        Origins of Financial Crises, The 175
credit crunch (see also: inflation; interest rates) 3, 17, 167-8, 223
Credit Suisse 246
        Credit Suisse Commodities Benchmark 246-7
crude
        benchmark varieties 52
        black gold 7-8
        'devil's excrement, the' 8
        industrial lifeblood 7
        fuel 8
        light, sweet (see: diesel)
        marginal barrel 28-30, 80-1, 248
        prime markers 52-3
        supply 23-8
Currie, Jeffrey 44-5, 146, 169-70, 231-3

D

D'Amato, Richard 38
Deffeyes, Kenneth (see: *Hubbert's Peak*)
Delta supply (see: Nigeria, supply)
demand growth (see: Asia, demand growth; crude, industrial lifeblood; supply and demand)
Department of Agriculture (US) 129
Department of Energy (US) 91, 129, 143-5
Department of Treasury (US) 129
derivatives 117-9, 121 (see also: commodities; commodity index investment; futures; options; swaps)
    investment 120
    off-exchange derivative agreements 120, 126, 221 (see also: speculation)
    OTC (over-the-counter) 103-5, 108-9, 115-21, 127-9, 137 188, 227   (see also: paper barrels)
developing world 3, 8, 21, 44, 83, 222, 234
diesel 144-5, 158-60, 189-91, 217
    crack spread 158
    demand 47
diesel fundamentalists 146, 158, 189
Diwan, Roger 83, 89, 93-5, 185-6
dotcom crash 5, 210-2 (see also: Nasdaq 100; "Oil dotcom")
Dow Jones-AIG (DJ-AIG) 114
Doyle, Sir Arthur Conan 77
Dresdner Kleinwort (see: Smith, Colin)
Drollas, Leo ix, 190-1, 215-8
Dubai (see: crude, prime markers)
Dunn, Michael 236

E

economic crisis 3, 10, 17, 223, 233, 235
economic growth 21-2, 32, 244
Economist, The 44, 134, 222

efficient market hypothesis 137-9, 171-4 (see also: speculation, opposition)
EIA (Energy Information Administration) 22-3, 91
    Caruso, Guy 91, 100
Emerging Asia 32-3, 37 (see also: Asia)
energy policy (see: resource nationalism)
Energy Security Analysis 83
ETF Securities 245
ethanol 25, 35
exchange-traded funds (ETFs) 108, 115, 117, 245 (see also: commodity index investment)
ExxonMobil 39

F

Federal Reserve (US) 129, 166, 171
feedback loops 7, 169, 174-83, 187, 222 (see also: "Minsky moments"; bubbles, negative bubble theory)
FERC (Federal Energy Regulatory Commission) 91
financial instability theory (see: Minsky, Hyman)
financialisation 9-10, 95-6, 251-2
    of oil 92-4
Financial Times, The 17, 133, 166, 227
    Alphaville blog 232
    "LEX" 134
floor brokers (see: speculation, floor brokers)
Forbes 229-30
foreign policy 30, 226 (see also: resource nationalism; Addax)
fuel 9 (see also: diesel)
    government subsidy 44
futures (see also: options)
    additive rules of open interest 71, 186
    delivery date 56
    front-month 41, 55-7 (see also: futures curve; oil price, Nymex price)
        "fully collateralised" 60-5

Nymex 54
offsetting 60, 63-6, 74
origins 54
oil 70-1
risk hedging 57-9, 63-6, 106, 119
speculation 66-7, 74-6, 135
spot delivery 55-6
futures curve 55, 102-4, 111-2, 149-51, 155, 188, 206, 221 (see also: speculation; swap dealers)
backwardation 111-2, 138, 155
contango 111-2, 138, 142, 155-7
"risk premiums" 207-8

G

GAO (Government Accountability Office) (US) 91, 192, 197
Issues Involving the Use of the Futures Market to Invest in Commodity Indexes 192
Gazprom Neft 226
Gensler, Gary 237, 241 (see also: CFTC)
Georgia (see: Russo-Georgian conflict)
Gibbs, Robert 10
Global Energy Studies (see: Drollas, Leo)
globalisation 22 (see also: Emerging Asia)
Goldman Sachs 44-5, 106, 121, 146-9, 227, 232-3, 243
2008 note 154-5
forecasts 243
Goldman Sachs Commodity Index (GSCI) 113-4
J. Aron & Co. 227
swap dealer 227-31
government fuel subsidy (see: fuel, government subsidy)
Great Crash 1929 169
Greenspan, Alan 171-2, 180
Gulf of Guinea 27
Gulf of Mexico 27

H

Hayward, Tony 247-8, 250 (see also: BP)
hedge funds 96, 108, 122
    market share 125 (see also: commodities, investment)
Hofmeister, John 82 (see also: Royal Dutch Shell)
Horsnell, Paul (see: Barclays Capital)
Hubard, William 107, 116
Hubbert, M. King (see: Hubbert's Peak)
Hubbert's Peak 34-5
    Hubbert's Peak, Deffeyes, Kenneth 35
Huntingdon prospect 224

I

IEA (International Energy Agency) 23, 39, 80
    Medium Term Oil Market Report 161
India 21, 44 (see also: Asia; Emerging Asia; resource nationalism)
industrialisation 21-2
Interagency Task Force (see: CFTC, Interagency Task Force)
interest rates 3, 16, 171, 223 (see also: credit crunch)
Interim Report (see: CFTC, Inter-Agency Task Force, *Interim Report*)
inventory levels 133 (see also: oil price, inventory levels)
    hoarding 140-2, 159-60, 190-1
    over-supplied 166
investor diversification 95
Iran
    geopolitical tension 31, 167
    nuclear program 31
Iraq
    invasion 30
    Kurdish oil licence 226
    production 43
    Turkish incursion 43
Irrational Exuberance (see: Shiller, Robert, *Irrational Exuberance*)

"Is it a bubble?" 157
Islamists 31

J

J. Aron & Co (see: Goldman Sachs, J. Aron & Co)

K

Keynes, John Maynard 173-4
Khelil, Chakib 47, 77, 94
Kirkuk pipeline (see: Iraq, Turkish incursion)
Kirsch, David 93
Krapels, Edward 83
Kurdistan (see: Iraq)

L

Laden, Osama bin 31 (see also: Islamists)
LCM Commodities 245 (see also: Morse, Dr Edward)
    "South by Southeast" 184
Lehman Brothers 46, 155-7 (see also: Morse, Dr Edward)
    bankruptcy 167, 169-70
"LEX" (see: *Financial Times, The*, "LEX")
light sweet crude (see: diesel)
long-dated oil (see: supply, fears)
Lowe, John 82
Lukken, Walt 100, 129-30, 192, 236 (see also: CFTC)

M

Mackay, Charles 15
Macquarie (see: Barakat, Nauman)
Malaysia 44
Malone, Robert 82
marginal barrel (see: crude, marginal barrel)
marginal cost of supply (see: crude, marginal barrel)

marginal production (see: Angolan deepwater; Canadian oil sands)
market makers (see: speculation, market makers)
market participants 193-5
    commercial/non-commercial split (see: speculation)
    criminality 205-6 (see also: speculation, speculators)
    market share 125
    pivotal trader 206
    swap dealers 221 (see also: swap dealers)
mark-to-market 175-6, 180, 184
Masters, Michael W. 51, 69, 76, 83, 126-8, 145 (see also: *Accidental Hunt Brothers, The*)
McCullough Jr., Robert 204-8, 222
Medium Term Oil Market Report (see: IEA, *Medium Term Oil Market Report*)
MEND (Movement for the Emancipation of the Niger Delta) 30, 43
Merrill Lynch 45, 216 (see also: Blanch, Francisco)
Mexico (see: Gulf of Mexico)
Middle East (see also: Iran; Iraq; OPEC)
Minsky, Hyman 6, 174-7, 179, 221
    "Minsky moments" 179-80
    Stabilizing an Unstable Economy 175, 177
Mississippi Scheme 7
MMTs (Managed Money Traders) (see: hedge funds)
Morgan Stanley 121, 227
Morse, Dr Edward 46, 133, 155-8, 160, 165, 184, 190-1, 200, 245-6, 248, 250
Murti, Arjun 44-5, 231

N

Nasdaq 100 5, 210-12
Nations Energy 38
negative bubbles (see: bubbles, negative bubbles)
'New Era' thinking 181, 183, 222 (see also: Shiller, Robert)
New York magazine 17
New York Post 18

Nigeria 8 (see also: MEND; resource nationalism)
    Bonga shutdown 168
    production 43
    supply 30
Norrish, Kevin (see: Barclays Capital)
Northern Rock 178
North Sea oil 27
Northwest Airlines 76
Nymex open interest (see: speculation, Nymex open interest)

O

Obama, Barack 10, 237
offsetting (see: futures, offsetting)
"Oil dotcom" 133, 156-7, 160, 165 (see also: Morse, Dr Edward)
Oilexco 224
oil price (see also: speculation)
    $100 oil 15-20 (see also: Arens, Richard)
    1980 record 20
    1990s/00s ascent 21
    ascent 41-2
    bust 165-70, 184 (see also: negative bubble)
    dark matter 117-120
    dotcom parallel 211-212 (see also: "Oil dotcom")
    fair price 245-51
    forecast 45, 81-3, 246
    front-month 68, 209
    global benchmark (see: oil price, Nymex price)
    industry cost 80
    inventory levels 78-80
    Nymex price 53, 68-9, 72
    speculators 188
    spot price 53, 65, 69, 76, 137, 146-7
    supply and demand 78
Olympics (see: Beijing Olympics)
Omimex 38

OPEC 9, 23-7
    non-OPEC members 27
    non-OPEC production 27, 29 (see also: marginal production; Russia)
    non-OPEC supply 32, 36-7
    production 29, 47, 244
    self-interest 251
    spare capacity 26-7
    speculation 94
    supply 30, 32, 36-7, 40, 217
open interest (see: Nymex open interest)
Oppenheimer & Co 83
options 73 (see also: derivatives; futures)
    options derivative writers 144-5
    speculators 158
Orinoco belt (see: Venezuela, Orinoco belt)
OTC (over-the-counter) derivatives (see: derivatives, OTC)

P

paper barrels 54, 70-2, 75, 137, 185-6
peak oil 34-5, 39-40, 147 (see also: supply, fears)
"perfect contango" (see: futures curve, contango)
Petro-Canada 38
PetroKazakhstan 38
PFC Energy 83 (see also: Diwan, Roger)
Phibro 106
pipelines (see: BTC; Kirkuk)
pivotal trader (see: market participants, pivotal trader)
Ponzi finance 6, 177-9, 222 (see also: Minsky, Hyman)
PPI index 248-50
Premier Oil 224
private sector (see also: resource nationalism)
production 40, 43, 47, 168, 217, 220, 244

R

real estate boom-and-bust 11
recession 8, 26, 179 (see also: credit crunch; economic crisis; interest rates)
regulation 10-11, 247 (see also: CFTC)
renewable energy 25, 36
resource curse 8 (see also: MEND)
resource nationalism 36-9, 42, 226
returns (see: speculation, commodities, returns)
risk hedging (see: futures, risk hedging)
"risk premiums" (see: futures curve, "risk premiums")
Royal Dutch Shell 39, 82, 224-6
Ruehl, Christof 77, 216-7, 220, 244, 247-8 (see also: BP)
Russia 37 (see also: resource nationalism)
     production 27-8, 39
Russo-Georgian conflict 168

S

Saudi Arabia (see also: OPEC)
     Al Saud dynasty, American support of 25
     crude capacity 27
     "oil shock" 9
     production 40, 47, 168, 217, 220
     spare capacity 160
     supply 39-40
     surplus production capacity 26
     terror attacks 31
Schork, Stephen 15, 17-20, 190-1
     Schork Report, The 15
Securities and Exchange Commission (SEC) 129
SemGroup 229-230
Senate Homeland Security and Governmental Affairs Committee (US) 127
Senate Subcommittee on Financial Services (US) 237
Shell (see: Royal Dutch Shell)

Shiller, Robert 6, 169-70, 175, 177, 181, 221
    Irrational Exuberance 4, 175
Simmons, Matthew R.
    Twilight in the Desert 40
Sichuan earthquake (see: China, Sichuan earthquake)
Sinopec 38, 226
Smith, Colin 46, 165, 246
South Sea Bubble 7
Special Call (see: CFTC, Special Call)
speculation 4-7, 55, 90, 94-5, 218
    blind herding 206-7
    bubble 7, 135-6
        definition 171-3
        oil price 187 (see also: oil price)
    commercial/non-commercial split 98-100, 102, 104, 106-7, 121-2, 125-6, 191-3, 206
    commodities
        indices 114-7, 127-8
        returns 109-14
    floor brokers 96
    futures 66-8, 74-6, 135-6
    future, the 251-2
    growth in speculative interest 95-6
    hedge funds 96
    market makers 67-8, 96, 125
    mechanism 218-9 (see also: swap dealer loophole)
    Nymex open interest 70, 125, 127-8, 185-7
    opposition 134-54, 161, 189, 216-7
        (see also: efficient market hypothesis)
    proponents 46-7
    speculators 83, 158, 188 (see also: swap dealers)
    spot forward gambit 206
    time spread trades 97
speculative finance 177 (see also: Ponzi finance)
spot price (see: oil price, spot price)

Sri Lanka 44
Stabilizing an Unstable Economy (see: Minsky, Hyman)
"stagflation" 9
Staff Report (see: CFTC, *Staff Report*)
Statistical Review of World Energy (see: BP, *Statistical Review of World Energy*)
Steenland, Douglas 76
stop order 19-20
Strategic Petroleum Reserve (SPR) 143-5, 191
structured investment products 108 (see also: commodity index investment)
Stupak, Bart 228
sub-prime (see: real estate boom-and-bust; Ponzi finance)
supply (see: OPEC; non-OPEC; marginal barrel)
    fears 147-154, 157
    shortage 190-1
supply and demand 9, 21-3, 77-8, 218 (see also: diesel fundamentalists)
swap dealers 102-6, 122, 125-7, 152-5, 194 (see also: CFTC; speculation, commercial/non-commercial split; Goldman, swap dealer)
    swap dealer loophole 102-5, 119, 126, 221, 233, 238
swaps (see: commodities, swaps; swap dealers)

T

Taiwan 44
Teitelman, Robert 182-3, 209
terrorism 31
TotalFinaElf 38-9
toxic assets 11
traders 125
tulipomania 7, 15
Turkey (see: BTC; Iraq, Turkish incursion)

U

Udmurtneft 38
United States-China Economic and Security Review Commission 38
United States Oil Fund 245

V

Venezuela
    Orinoco belt 37
    resource nationalism 39
Verleger, Philip K. 126, 143-5, 191, 204
Vitol 202-3, 206

W

War on Terror 31
Washington Post, The 228
West African oil 27
Wray, L. Randall 226
    Commodities Market Bubble, The 223
WTI (see: crude, prime markers)

Y

Yeats, W. B. 243
Yukos 37